Computational Thinking

Hideyuki Nakashima · Keiji Hirata
Editors

# Computational Thinking

## Rethinking How We Think

Springer

*Editors*
Hideyuki Nakashima
Sapporo City University
Sapporo, Hokkaido, Japan

Keiji Hirata
Department of Complex and Intelligent Systems
Future University Hakodate
Hakodate, Hokkaido, Japan

ISBN 978-981-95-5961-9 ISBN 978-981-95-5962-6 (eBook)
https://doi.org/10.1007/978-981-95-5962-6



The original submitted manuscript has been translated into English. The translation was done using artificial intelligence. A subsequent revision was performed by the author(s) to further refine the work and to ensure that the translation is appropriate concerning content and scientific correctness. It may, however, read stylistically different from a conventional translation.



This Springer imprint is published by the registered company Springer Nature Singapore Pte Ltd.
The registered company address is: 152 Beach Road, #21-01/04 Gateway East, Singapore 189721, Singapore

# Preface

When I first read Jeannette Wing's influential essay "Computational Thinking" published in 2006, I immediately recognized its importance. I wholeheartedly agreed with her assertion that computational thinking (CT) should be a fundamental component of university education for all disciplines, and that students majoring in computer science would be equipped to excel in a wide range of careers beyond their technical training. I translated Wing's essay into Japanese, and it was published in the Journal of the Information Processing Society of Japan in 2015.

However, when Prof. Michael Vallance, our colleague at Future University Hakodate (hereafter referred to as FUN), used my translation as the teaching material, he found that students struggled to understand it. Wing's essay includes many technical terms from computer science but offers few concrete examples—particularly ones that demonstrate how such concepts apply to other fields. While the essay was written for educators, it is not structured to help beginners grasp the essence of computational thinking.

This observation led us at FUN to launch a project aimed at publishing a book filled with accessible and practical examples of CT. It was not easy to find examples that made sense to people outside of computer science. We each tried various approaches and held frequent discussions. Eventually, we compiled our findings into the desired book in Japanese, which translates Wing's abstract ideas into concrete terms that are easier to understand for non-specialists.

At the same time, international interest in teaching computational thinking was gaining momentum, with new academic societies and initiatives being launched. In Japan, however, government agencies such as the Ministry of Education had interpreted CT to mean programming and began to incorporate it into elementary school curricula. Even national media outlets like NHK produced TV programs based on this interpretation. Needless to say, programming knowledge and skills, along with the ability to effectively utilize modern AI technology, are certainly part of CT, but CT encompasses much more.

This English edition is based on the Japanese edition published by FUN Press in 2022. Chapters 1, 2, 4 and 5 of that Japanese edition have been revised to a diverse extent. This English edition was unable to include Chap. 3, authored by Michael Vallance. For his research, please see his paper titled "Computational thinking and interdisciplinary learning: time to focus on pedagogy" published in 2020.

The structure of this book is designed to be progressive, starting with an introduction to fundamental concepts in computer science, moving through methods for understanding computational thinking, and culminating in its practical application. However, it has been consciously written so that readers can begin reading from any chapter.

This book begins in Chap. 1 by serving as a primer for readers unfamiliar with computer science, defining computational thinking as "thinking like a computer scientist." The chapter explains the origins of computer science and its unique worldview by contrasting it with fields like physics and mathematics, while also addressing its relationship with programming, deep learning, and generative AI.

Chapter 2 delves into the practical benefits of applying computational thinking in the real world. It introduces three of four essential concepts—abstraction, modeling, and virtualization—which are presented as fundamental operations for generating new information. The chapter uses relatable examples to make these terms accessible and concludes by connecting computational thinking to the principles of information design.

Chapter 3 introduces the fourth essential concept, meta, and its related idea of self-reference. Acknowledging the complexity of the topic, the chapter explains how to think about an object and concept from a higher perspective, using real world examples to clarify how to apply this thinking both in human activities and within a computer.

Chapter 4 is an essay in which a computer scientist and education scholar shares her personal journey, explaining how her childhood experiences cultivated a computational mindset. She uses the unique analogy of cooking to explain computational thinking, breaking down a recipe into a series of logical steps, and discusses how this perspective can be taught effectively to children.

Finally, in Chap. 5, a cognitive scientist with no background in computer science reports on her personal struggle to understand computational thinking. She shares her initial discomfort and the turning point that allowed her to grasp the concept, concluding with a non-specialist's unique discovery of computational thinking's altruistic power.

Sapporo, Japan
September 2025

Hideyuki Nakashima

# Contents

# Chapter 1
# What is Computational Thinking?—Think as Computer Scientists Do

**Hideyuki Nakashima and Keiji Hirata**

**Abstract** This chapter is written to provide foundational knowledge about computers for readers unfamiliar with computer science, serving as a primer. Jeannette Wing, who proposed computational thinking, defined it as "thinking like a computer scientist." Computer science is a discipline that seeks to understand the nature of the world based on the worldview of information and computation. Computer science is a relatively young field, having only begun to be recognized as an academic discipline around the mid-twentieth century. This chapter explains the origins and concepts of computer science, detailing unique ideas not found in other academic fields, while carefully guiding beginners to understanding. This chapter provides a clear definition of computational thinking by examining the thought processes of computer scientists. It includes a wealth of concrete examples to enhance understanding. A comprehensive perspective is offered by contrasting the meaning of computation with concepts from physics, logic, philosophy, and mathematics. It also addresses the relationship between programming and computational thinking, which is often confused, as well as its connections to deep learning and generative AI.

## 1.1 Thinking in the Computer Age

Without computers, we would not be able to use electricity, water, banks, railways, or airplanes. Understanding how computers accomplish those tasks also helps us think and plan our daily life tasks. Computational thinking, which this book deals with, is not just knowledge or technology, but something like a compass for living better in the information world.

H. Nakashima (✉) · K. Hirata
Sapporo City University, Sapporo, Hokkaido, Japan
e-mail: hideyuki.nakashima@gmail.com



H. Nakashima and K. Hirata (eds.), *Computational Thinking*,
https://doi.org/10.1007/978-981-95-5962-6_1

## 1.2 Computation and Thinking

The term computational thinking contains two words that everyone knows. One is computation, and the other is thinking.

### *1.2.1 Computation*

When you hear the word "computation", you might think of arithmetic operations. However, computations in the information world are not limited to arithmetic operations. Thinking of tactics to win a soccer game, planning a 10-day vacation in Vienna, creating the best dish with the ingredients in the refrigerator, ensuring logistics when roads are cut off by an earthquake, and even writing novels or composing music are all computations. What these have in common is that they are about "finding the optimal solution" for a goal within the constraints of the real world.

Computation is a powerful method for generating new information. For example, finding the greatest common divisor, GCD, of two numbers is not an easy task in general. The GCD of 4 and 6 can be easily found (2), but finding the GCD of large numbers is difficult. There is a method of Euclid for finding the greatest common divisor GCD of two numbers (defined in Sect. 1.7.5), which is generally referred to as "Euclidean algorithm". This kind of computational method is called an "algorithm."

You may ask what does finding GCD have something to do with our daily life. For instance, one textbook shows an example of a task of tile craftsmen.

You are trying to cover the floor of a room with short sides of 18.9 m and long sides of 24.5 m with square tiles. You are about to bake the tiles, but you want to use as few tiles as possible. How long should one side of the tile be?

The correct answer is 70 cm, as the GCD of 189 and 245 is 7. Note that the GCD is a concept that applies only to natural numbers, so we are multiplying by 10 for the calculation.

In actual daily life, you may rarely see situations where the GCD is used. However, the expression "greatest common divisor" is often used metaphorically in everyday life. For example, it means the largest common element of two or more opinions, which in this case means the compromise point of the discussion.

On the other hand, there is also the least common multiple. This means the smallest multiple of two natural numbers. There is a zodiac in the calendar, which consists of the combination of ten stems: 甲 (Jia), 乙 (Yi), 丙 (Bing), 丁 (Ding), 戊 (Wu), 己 (Ji), 庚 (Geng), 辛 (Xin), 壬 (Ren) and 癸 (Gui); and another twelve stems: 子 (rat), 丑 (ox), 寅 (tiger), 卯 (rabbit), 辰 (dragon), 巳 (snake), 午 (horse), 未 (goat), 申 (monkey), 酉 (rooster), 戌 (dog) and 亥 (pig). Each year is named by the combination of two stems. The least common multiple of 10 and 12 is 60, so the zodiac returns to its original state every 60 years. This is called the sexagenary cycle.

### *1.2.2 Thinking*

When you hear the word thinking, you might think it's all about mulling over ideas in your head. However, the thinking in computational thinking doesn't stop there. What's important is to actually try out what you've thought about in the real world, in other words, practice (more about practice will be discussed in Chap. 2). Imagine yourself as the hardware and software of a computer, freely executing computations in the real world surrounding you. Jeannette Wing, who proposed computational thinking, has stated that computational thinking is thinking like a computer scientist [Wing 2006]. As you will see as you read this book, computer scientists use the tool of the computer, or the metaphor of the computer, to try to understand various phenomena and problems in the real world.

In the real world, everything exists, functions, and changes under the absolute constraints of physical laws. Specific constraints include time progresses, entropy increases, and energy and mass are conserved. Neither the human body, the earth, nor space rockets can escape these physical constraints. On the other hand, computations are generally abstract operations and are not subject to physical constraints. In the physical world, there are physical law constraints to build a building above a certain height, and there is a limit to the height. However, there are no such constraints in the world of computation.

For example, no matter how long the number of digits, it is possible to perform arithmetic operations, and no matter how large the database, it can be created if desired. However, it is not possible to actually prepare an infinite-digit abacus (or computer memory) to hold the numbers to be calculated, nor is it possible to actually create a database with a record number exceeding the number of atoms in the entire universe. Therefore, at least in the real world, it is meaningless unless arithmetic operations can be performed, and databases can operate within the realistic range of costs, time, and space. Although it is not constrained by physical laws, there are constraints in the world of computation, such as the amount of memory and computation time (collectively referred to as "computational complexity"). This is an important concept in computational thinking.

### *1.2.3 Conditions for Good Computation*

Furthermore, the computation in the real world inevitably involves the perspective of value. It is because of the physical constraints mentioned in the previous section that value becomes necessary. For example, one person may prioritize the speed of arithmetic operations over the size of the number of digits. Another person may prioritize the number of queries that can be processed at once over the response time for each query to the database. This means that with the introduction of value, computations are classified into desirable and undesirable ones, and if that value changes, the desirable computations will also change.

**Column 1 The difficulty of value evaluation**

You may often want to optimize the procedure of your work. It could be the procedure of cooking or cleaning, but "what to optimize?" is a difficult question. For example, it is difficult to create a measure for "taste." It also depends on individual preferences (values).

Even if a measure can be defined, there is a problem of trade-offs between multiple measures. In cooking, for example, there are various measures such as cost of ingredients, cooking time, taste, etc., and they often contradict each other. If you try to optimize only the taste, the cost of ingredients will be higher and the cooking effort will increase. Humans are making compromises between these multiple conditions that cannot be optimized at the same time. For example, optimizing the taste within a cost of 1000 yen for ingredients.

When a JR (Japan Railways) express train runs late, it's standard practice to hold local trains to keep the express on schedule. From JR's point of view, the express train's punctuality is a higher priority because delays beyond a specific duration often result in ticket refunds. As a passenger, you always want your own train to have the highest priority, but that's not always the reality. One of the authors once learned this the hard way on the Airport train heading from Sapporo to New Chitose Airport. An announcement came on: the Airport train would stop temporarily to let a delayed express train pass. Despite having left ample time, the author missed their flight. In the moment, the author felt the missed connection was simply a consequence of JR's scheduling priorities. Yet, when considering the severe disruption the flight connection caused—and not just for one person—the importance of airline connections should, arguably, be given higher weight. Of course, everyone's situation is unique, and not all priorities align. Ultimately, this scenario highlights a conflict between JR's internal operational values and the values of time-sensitive passengers.

Currently, corrections to schedule disruptions due to accidents or bad weather are mostly done manually. Optimization is something computers are good at, but it seems that it has not been programmed because it is not self-evident which conditions should be optimized (they change depending on the situation). At this point, it is unclear whether AI will be able to understand the values held by humans in the future.

A good computation in computational thinking gives the optimal solution from a certain value perspective to the goals set within such physical constraints. The result changes depending on which of the multiple values is chosen. Computational thinking is a way of thinking that provides shortcuts and hints to realize this good computation.

No matter how smart a robot is, or how well an autonomous vehicle drives compared to a human, or an AI that creates works of art, the computer is performing symbol manipulation within it.[1] Through these manipulated symbols, the computer understands the real world and then influences it. What becomes important here is how to assign (correspond) symbols to the things and concepts of the real world bound by physical constraints. For example, the computation needed to answer the question "How many minutes are in a day?" is immediately clear. The meaning of the symbols "day", "hour", and "minute" that appear in that computation is obvious. However, symbol manipulation of "joy" or "sadness", symbols assigned to the movements of the mind, seems difficult or almost impossible. "Every time I listen to Bach's music, I discover something new", "I don't mind sacrificing myself for justice", is it possible to realize a computer that calculates such things? This is a fundamental problem of artificial intelligence. And it has been debated for a long time. Can intelligence be expressed with symbols alone?[2] Let's leave such difficult questions for another occasion. In this book, we only want to consider how to represent phenomena, constraints, and problems of the real world as data structures and functions in a program. Through this, we want to change our view of the world. As for how to take such perspectives, set problems, and educate them, Chap. 4 will also address this.

The issue of representation is a historical challenge in computer science research, and it has been dealt with in the artificial intelligence research from the early stages, but it has not been given much importance in traditional education. The first condition for good computation is appropriately representing real-world problems within a computer, and the same way of thinking can also be useful in our daily lives. I would like to provide examples of this. The understanding of the real world by a computer through symbols within a computer is quite similar to human's understanding of the real world through metaphors and analogies. Using metaphors and analogies allows us to feel as if we understand "it" by using similar parts as clues, but it does not accurately represent all of "it", nor does it mean that we have come to understand all of "it." However, good metaphors and analogies greatly help humans' understanding and are very useful.

The second condition for good computation is to explore and find more efficient or effective ways to solve phenomena or problems that have been brought into the world of symbols. Algorithms and procedures are ways to show the path to solutions. They can be used not only for computer programs but also for humans. Reducing the size of the array needed for computation or reducing the number of iterations is involved in efficient and effective problem-solving. This way of thinking is effectively applied to human thought as well, probably because human memory capacity, human recognition/computation speed and human error rates are also influenced by physical constraints.

Here we summarize the conditions of good computation; the first condition is to appropriately represent real-world problems within a computer, and the second one is to explore and find more efficient and effective ways to solve phenomena or

[1] When we were writing the original Japanese version of this book in 2021, this statement was true. However, ChatGPT released in late 2022 changed the scenery.

[2] Deep learning and generative AI do not use symbol manipulation in the sense written here; they operate using high-dimensional vectors instead.

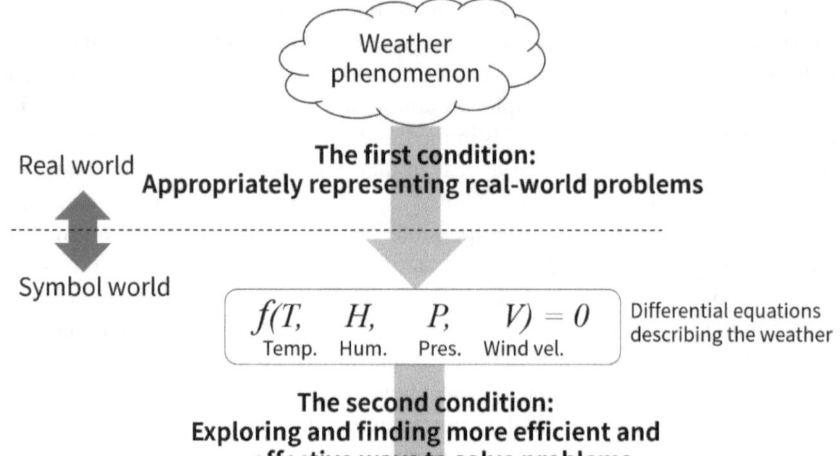


**Fig. 1.1** Computational thinking that satisfies the two conditions for good computation

problems that have been brought into the world of symbols. Let's illustrate the relationship between these conditions for good computation and computational thinking (Fig. 1.1).

When facing phenomena and problems in the real world that seem ambiguous, massive, and sometimes even structureless, how can we effectively meet these two conditions of good computation? Let's consider this problem in detail below.

## 1.3 Exponential Explosion

When programming a computer for calculation, one must be careful of a phenomenon known as exponential explosion. This refers to the situation where the amount of data increases and the computation time and memory capacity required increase too much to handle. Such a situation is called the exponential increase as mentioned in the column, and "explosion" is a term that emphasizes the point of rapidly becoming intractable.

For instance, exponential increase can be reproduced by a process of adding up all the numbers that have appeared so far to create the next number as follows.

$$X_{n+1} = \sum_{i=1,\ldots,n} X_i$$

In fact, let's see what happens if we start with $X_1 = 1$, and you obtain the sequence of 1, 1, 2, 4, 8, 16, 32, 64, … As is evident, starting from the second term, numbers grow exponentially and can be expressed as $2^n$. Why does the calculation produce such a beautiful sequence of numbers? You may understand it immediately if you think of the asymptotic formula to compute $X_{n+1}$ as follows.

$$\begin{aligned} X_{n+1} &= \sum\nolimits_{i=1,\dots,n} X_i \\ &= X_n + X_{n-1} + X_{n-2} + \cdots + X_1 \\ &= X_n + X_n \\ &= 2X_n \end{aligned}$$

### 1.3.1 Friends of Friends of Friends of…

There are many examples in the world where numbers increase exponentially. You may have heard the fun phrase "a friend's friend is a friend." If each person has 10 friends, then a simple calculation tells us that friends of a friend would amount to 100 people. This situation is expressed as having 10 friends at the first level and 100 friends at the second level, and thus, the number of friends increases exponentially. Now, in sociology, there is a well-known rule called "six degrees of separation." This means that anyone can become friends with all the people on earth if he or she considers up to the sixth level of friends (assuming there is no overlap in friendships). The current global population is 7 billion, which is approximately $44^6$; this implies that each person has an average of 44 friends. Intuitively, it seems reasonable that each of us has about 50 to 100 friends or acquaintances.[3] What is surprising here is that by following the chain of friends-of-friend only 6 times, you can reach everyone on Earth. Amazingly every reader of this book can reach the President of the United States by tracing your friends-of-friend chain for 6 times! You might feel the power of exponential growth.

**Computational complexity**

In the world of programming, the effort required for (the memory and computation time required to execute the computation) is an issue. Th is shown as a quantity relative to the size of the problem and is called the order, which is represented by a function *O*. For example, *O(N)* means that resources proportional

[3] The cognitive limit for the number of stable, meaningful social relationships an individual can maintain is estimated to be 150 people, known as Dunbar's number.

to the size of problem are needed. $O(k^N)$ is called exponential the computational complexity increases rapidly with the size of problem; it is usually not practical for computation. Conversely, $O(log\ N)$ is an easy problem to handle.

**Column 2 Exponential Acceleration of Evolution**

Both the evolution of organisms and the evolution of science and technology have been developed so far through the combination of "parts." Evolution accelerates exponentially as the parts get larger. In terms of organisms, the evolution that was occurring at the molecular level extends to the evolution at the cellular level due to the creation of a part called a cell, and it became dramatically faster (incidentally, it is said that all existing organisms are descendants of a single cell that was created by chance). Furthermore, multicellular organisms were born from unicellular organisms, giving birth to a new unit of evolution. What was once a whole is divided into parts, and these individual parts are differently combined to create a new whole, which becomes a new unit. It gives rise to the evolution at a higher speed.

Then humanity was born and gave birth to science and technology. The realm of the evolution shifted from the level of individual organisms' intelligence to the that of the intelligence of the entire society created by those organisms (humans). Necessarily, the laws governing evolution were also changed. Until cells were created, chemical reactions determining molecular structures were the dominant laws. When a cell became a unit, the dynamic non-equilibrium system as life became the governing principle. When organisms became a unit, the theory of evolution in the sense described by Darwin became the governing principle. Furthermore, when the knowledge created by humanity became a unit, the methodology of science and technology became the governing principle.

Science and technology have been developed on the sum of past knowledge. We learn only in a few hours the knowledge that old scientists like Galileo and Newton discovered over tens of years. Then we connect it to the next discovery. Thus, human knowledge is also increasing exponentially.

If we look at the evolution of evolution in a timeline, we should recognize that the evolution of evolution is being accelerated exponentially:

| | |
|---|---|
| 4.5 Billion years ago: | Birth of the Earth |
| 4 billion years ago: | Birth of life. |
| 2 billion years ago: | Birth of cells. |
| 1 billion years ago: | Birth of multicellular organisms. |
| 0.5 billion years ago: | Birth of eyes (Cambrian explosion). |
| 0.07 billion years ago: | Birth of primates. |
| 0.005 billion years ago: | Birth of humanity. |

| | |
|---|---|
| 0.00001 billion years ago: | Birth of agriculture. |
| 0.0000003 billion years ago: | Industrial revolution. |
| 0.00000007 billion years ago: | Birth of computers. |

### *1.3.2 Cell Division*

Let's consider the number of cells as an example of a living thing that increases exponentially. You know that organisms grow by cell division. Imagine a fictional creature whose every cell doubles every day. If this organism was 10 cm tall one day, please guess how tall it would be the next day. No, it's not 20 cm. The correct answer is that the height the next day is the cube root of 2 multiplied by 10 cm. Doubling the height means doubling the width and depth, so the volume increases eightfold (Fig. 1.2). Going back to the first question, if the number of cells doubles in a day, how many days will it take for the height to double? Yes, it's for 3 days. When the height doubles, the weight becomes 8 times. It takes 1 day for the weight to double, 2 days to quadruple, and 3 days to increase eightfold. If the fictional creature grows at this rate, it takes only 27 days for the organism that was 10 cm on the first day to become 50 m tall, which is the height of Godzilla. This is because 50 m is 500 times 10 cm (here, we use a commonly used approximation $500 \approx 512 = 2^9$), so the volume is approximately $2^{27}$ times. It is not surprising that an organism of only 10 cm in height grows into Godzilla in 27 days by cell dividing once a day. This is the power of exponential increase.

### *1.3.3 The Number of Your Ancestors*

Every human has two parents, and four grandparents. That is, you have two ancestors one generation back, and four ancestors two generations back. In this way, the number of your ancestors increases exponentially. So, how many ancestors would you have if you go back 30 generations? It's about 1 billion. If we assume one generation is about 30 years, 30 generations is 900 years. So, you were born from about 1 billion ancestors who lived on this Earth 900 years ago.

But you may feel something is wrong here. The actual world population 900 years ago was just over 300 million. Generally, when a couple gets married, the husband has four grandparents, and the wife has four grandparents. But if cousins marry, the total number of grandparents is reduced to six. That is, if you trace your paternal and maternal ancestors, there must be some overlap. In other words, your father and mother can be traced back to one common ancestor at some point. In this sense, almost all people are related by blood somewhere.

**Fig. 1.2** Doubling the height results in an eightfold increase in volume

### *1.3.4 Pyramid Scheme*

Let's explain exponential growth with an example of a pyramid scheme or multi-level marketing. The job of a pyramid-scheme member is to recruit new members and to force them to purchase and sell products. If you recruit someone and they become a member, that person becomes your direct downline (child member). That is, you are the parent member of that new member. Then, a new member purchases a product from you and sells it to his/her child member with a certain amount of commission fee added. According to the rule of the multi-level marketing, a member who earns a commission fee must pay a certain percentage (pay-up rate) of the commission fee to the parent member.

Each member has a mission to recruit more than N new members, and there is a quota for product procurement and sales. People who are invited to become a pyramid-scheme member by an acquaintance may hesitate whether to become a member or not, but then the parent member says something like "The earlier you become a member, the more income you will earn," and urges them to join.

Is it true that "the earlier you become a member, the more income you will earn"? Let's actually calculate it and check with concrete figures. Suppose that it is a rule that one member recruits at least two new members, purchases a product for 100 yen, sells it for 300 yen, and the quota is 100 items. Thus, the commission is 200 yen per item, for a total income of 20,000 yen. If you pay 50% (pay-up rate) of the commission fee to the parent member, you will have 10,000 yen left, and you will

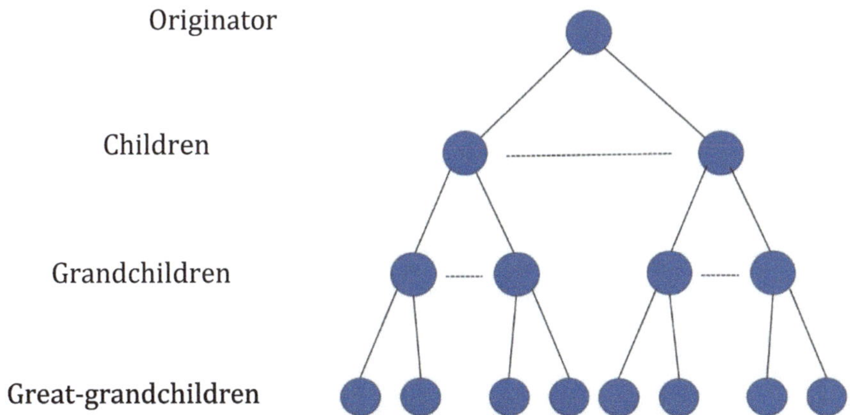


**Fig. 1.3** Structure of generations in pyramid scheme

pay 10,000 yen to the parent member. Since that parent member has at least two child members, the total is two-fold 10,000 yen paid up. If there is a parent member above that parent member, you must pay up 10,000 yen, which is 50% of 20,000 yen, but if you think that the child member earned 10,000 yen while the parent member was sleeping, it's a super easy business! Conversely, let's consider the case where you have grandchild members. Suppose you have two grandchild members for each of the two child members, then a total of 40,000 yen is paid up from the four grandchild members to the two child members, and 20,000 yen is paid up from the two child members to you, and you pay up 10,000 yen to your parent member. In other words, you have indirectly received 10,000 yen from the four grandchild members. As you can see from a little calculation, as long as there are great-grandchildren and great-great-grandchildren beyond the grandchild members, you will receive 10,000 yen for each generation increase (Fig. 1.3).

Now, let's get to the point. Let's reveal the answer to the question, "Is it true that the earlier you become a member, the more income you will earn?" The answer is qualitatively correct, but if you apply specific numbers and calculate, you will find that it breaks down immediately. The reason why this pyramid scheme works is the increase in members, but the speed of that increase is the problem. The mechanism of the pyramid scheme, "one member enrolls N or more new members," leads to an exponential increase in the number of members. $N = 2$ is the so-called doubling game, which doubles in order: 1, 2, 4, 8, 16, 32. At first, these are small numbers, but at the 10th step, it's about one thousand,[4] at the 20th step, it's about 1 million, and at the 30th step, it's about 1 billion, rapidly inflating. In other words, this pyramid scheme reaches almost the entire population of the earth after 30 generations, and that's where it ends. Under these conditions, the "originator" who started this pyramid scheme can only obtain a mere 300,000 yen.

[4] To be precise, $2^{10}$ is 1024, but remembering that it is about 1000 is useful for various calculations. Even larger numbers such as one million and one billion can also be easily calculated in your head.

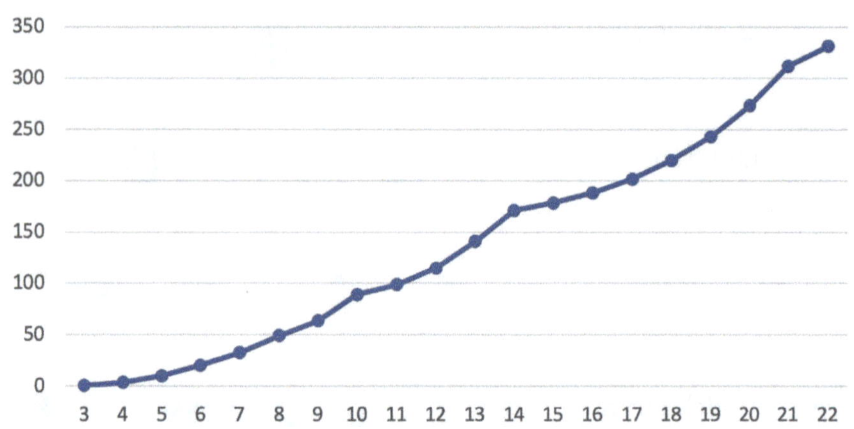


**Fig. 1.4** Pyramid-scheme: number of child members and originator income (10 K yen) at pay-up rate of 50%

$N = 2$ shows that you can't generate much money. In a pyramid scheme, you must increase the proportion of the pay-up or increase the number of child members.

Figure 1.4 shows the increase in the number of child members in each generation. We increased it from 3 to 22. Although there is an increase approximately proportional to the number of child members, the maximum is only about 3.5 million yen, so it's not a big profit.

Figure 1.5 shows the change in the pay-up rate. We changed it from 0.1 to 1.0. Here, there is an exponential explosion of increase. A pay-up rate of 1.0 means that the parent takes all, so only the originator who started first gets the money, and there is no income for the other intermediate generations, so it does not work as a pyramid scheme. But as we saw above, at the pay-up rate of 0.5, it doesn't make much profit, so you might aim for 0.7 (income of 42 million yen) or 0.8 (income of 400 million yen). The closer the pay-up rate gets to 1.0, the more the originator's income increases. However, the rate of increase is difficult to discern in Fig. 1.5, so we introduce a logarithmic scale on the vertical axis. This is shown in Fig. 1.6. From 0.5 to the right, it's almost a straight line, so you can see that it shows an exponential increase.

Either way, when you calmly look at the figures relevant to the pyramid scheme shown here, it would seem odd that many people in the world are being duped by the pyramid scheme. There may not be anyone who profited a lot from the pyramid scheme. It is probably because people have a vague understanding of exponential increase and suppose that it is naively applied to their own profit in the pyramid scheme.

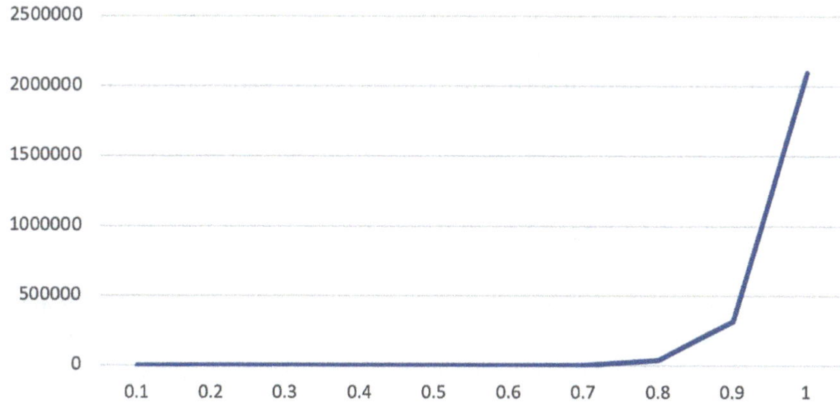


**Fig. 1.5** Pyramid-scheme: pay-up rate and originator income (10 K yen)

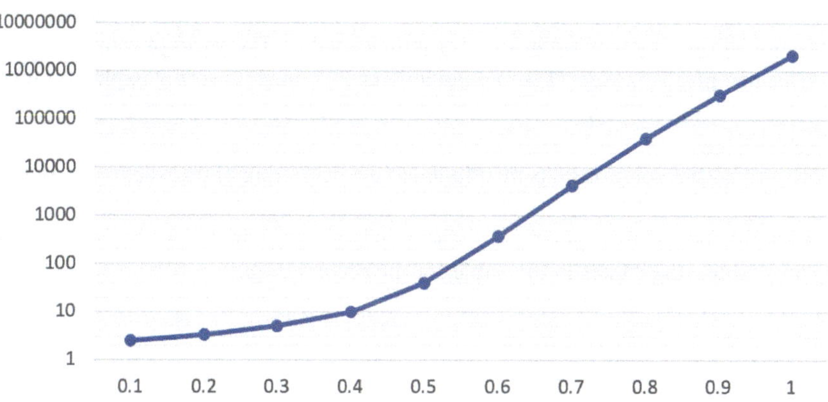


**Fig. 1.6** Pyramid-scheme: pay-up rate and originator income (Log scale in 10 K yen)

### *1.3.5 Traveling Salesman Problem*

Let's consider a more realistic example, there is a problem called the traveling salesman problem. This is a problem in which a salesman finds an efficient stroke order to travel multiple cities for sales activities, i.e., making the plan of a visit, stopping only once in one city with saving time and transportation costs. It has been known as a typical example of exploration for a long time and one of the most difficult problems to find the optimal solution. It is because the computational complexity increases exponentially as the number of cities visited grows.

Here, we will not give a solution to this problem but just an important notice thinking of actual logistics. For example, when a courier delivers multiple packages to multiple homes, in what order is it most efficient? In reality, there are other conditions that we have to take into account such as the specified time of delivery and the total mileage, which make the problem more complicated. Without a computer, the optimal solution cannot be obtained. From the standpoint of computational thinking, the important thing is not to solve this problem, but to know that this problem is difficult.

### *1.3.6 Avoiding Exponential Explosion*

As we have seen so far, when an exponential explosion occurs, the amount of computation increases rapidly, making it impossible to compute with practical computational resources (time and memory that are available). There are some known ways to avoid this:

- Divide and conquer: This is a method of dividing the problem into independent units and solving each one (Sect. 1.4).
- Settle for an approximate solution: This is a method of giving up on complete computation and accepting an approximate value. For example, the traveling salesman problem (Sect. 1.3.5) inevitably causes an exponential explosion when calculated normally, so several methods have been developed to find approximate solutions.
- Record partial solutions: As mentioned at the beginning of this section, exponential explosions often have a structure where you repeatedly use what you have accumulated so far, so you record what you have calculated in a table at each time step and use the values in this table for the second time.

## 1.4 Divide and Conquer

Previously, we discussed how an exponential explosion in computations can become unmanageable. Essentially, this explosion occurs because we have to consider all possible scenarios or cases of the subject. Therefore, let's first consider properly dividing the entire subject into small parts that do not cause an exponential explosion. If there is no relationship between these parts, or if any relationship can be ignored, the processing of each divided part will be light. This might allow us to calculate the entire subject without causing an exponential explosion. In other words, if we can divide it well, we can govern it with a light effort. In the best-case scenario, the entire computation can be completed with effort proportional to the number of divided parts. In other words, the original computational complexity of

$$N_1 \times N_2 \times \cdots \times N_k$$

can be reduced to only

$$N_1 + N_2 + \cdots + N_k.$$

When you hear divide and conquer, you might think of the ancient Roman Empire, which governed vast territories, Western colonial management, or the Edo Shogunate. The key to divide and conquer here was to sever the solidarity and cooperation between regions, ethnic groups, and countries (sometimes, the rulers even incited mutual resentment and hatred). By severing these, they were able to govern each divided region, ethnic group, and country with minimal effort in total.

## 1.5 Recursion

According to the Super Daijirin, a Japanese dictionary,

**Recursive**

1. The result of one's actions returns to oneself. Feedback.
2. In mathematics, etc., the thing to be defined appears in the definition itself.

The second definition is our concern here.

"Recursive" is one of the important and powerful methods in programming, which allows many problems to be described concisely by including calls to itself in the program. First, I would like to start with an example of what recursion is.

This phenomenon can also be seen in everyday life. For example, in a pair of mirrors facing each other, image B is reflected in A, and A is reflected in the image of B, and this continues indefinitely.

**Column 3 Mutual Exclusion**

If multiple processes access a single resource at the same time, one result may overwrite the other, causing problems. Mutual exclusion is to ensure that only one process can access at a time.

When accessing a single resource, you first lock the resource to prevent other processes from accessing it. This is called a lock. You can safely access the resource after locking it and unlock it when you're done. If another process wants to use the resource while it's locked, it has to wait until it unlocked. This is how mutual exclusion is achieved. This book is also written by multiple authors, so we use mutual exclusion when modifying files.

The key to efficient mutual exclusion is to keep the interval between lock and unlock as short as possible. To achieve this, we carefully consider what should be shared resources and devise algorithms accordingly.

Mutual exclusion is also used in railways. Nowadays it is computer-controlled, but in the past, to prevent more than one train from entering a single track, only one tablet was prepared for that section (called a block section), and trains without it could not enter. This was featured in the Japanese movie "Railroad Man" (original title "Poppoya") starring Takakura Ken.

Speaking of which, it seems that there are many good hackers among model railway enthusiasts. (Or should I say, there are many model railway enthusiasts among good hackers?).

There is a Rakugo (Japanese traditional storytelling) called "Hell's Eight Scenic Spots of the Dead". A Rakugo storyteller who goes to hell will be able to go to paradise if he makes Enma (hell's guardian) laugh, so he tells Enma a Rakugo story.

> Enma: Oh, a Rakugo storyteller, I love Rakugo. What kind of story is it? Rakugo storyteller: How about "Hell's Eight Scenic Spots of the Dead"? Enma: Oh, let's listen to it.

So, the same Rakugo starts in the Rakugo. Then, "Hell's Eight Scenic Spots of the Dead" starts again in that Rakugo, and it doesn't stop. In reality (?), however, Enma should stop it halfway through.

Describing the story in the order in which things happened like above, it continues to unfold indefinitely. However, if you think about it the other way around, you can describe repetitive tasks or computations in a short way. If you can capture various phenomena in everyday life recursively, you will be able to understand things in a simpler way. For example, the predicate "append" that connects two lists in the programming language Prolog is defined as follows:

```
append([], Z, Z).
append([E|X], Y, [E|Z]) :- append(X, Y, Z).
```

The first line is the termination condition, and computation stops when the first argument becomes an empty list. If the first argument is not empty, the control goes to the second line; it is decomposed into the head element E and the rest X, and the predicate is to append E and Z at the place of the third argument. Here, Z is the result of append of X and Y obtained from the recursive computation of append(X, Y, Z). It is enough to see that the second line is recursive. The LOGO language's "word list" program (Sect. 4.4) is also written recursively.

This is a strategy of dividing a large task or computation into smaller ones (see also Sect. 1.4).

A famous example where recursive solutions are effective is the Tower of Hanoi puzzle (Fig. 1.7). This involves moving stacked discs one at a time to move the entire pile to another rod. You must not stack a larger disc on top of a smaller one. This is easy to think about recursively (Fig. 1.8).

1. In order to move $N$ discs to the target rod
2. Move $N$–1 discs to a different rod named A (recursive procedure)
3. Move the remaining 1 disc to the target rod
4. Move the $N$–1 discs stacked on rod A to the target rod (recursive procedure).

In other words, to move $N$ discs, you need to use the operation of moving $N$–1 discs twice, in which the slightly smaller move operation is called recursively.

**Fig. 1.7** The Tower of Hanoi puzzle

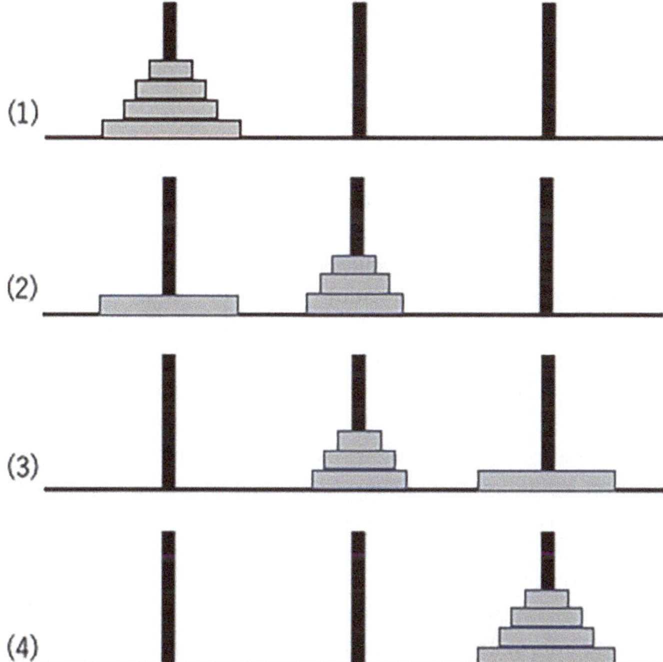


**Fig. 1.8** How to move discs in the Tower of Hanoi

## 1.6 Information as a Worldview

"Computational thinking" may be rephrased as "information theoretic thinking." This emphasizes information as a worldview alongside matter and energy. In other words, it advocates viewing the world from the perspective of information and its operations.

Have you ever read the flash fiction written by a Japanese novelist Shinichi Hoshi titled "The Net of Voices"? One of the characters claims that information is energy:

> In an era without knowledge of hydroelectric power, people probably didn't think of water as an energy source, and they certainly didn't dream that uranium-containing ore was a tremendous source of energy. Uranium ore was just a rock. Isn't it natural to think that the source of energy is the knowledge of nuclear power? In the early days of the nuclear age, information about atomic bombs was a high-level state secret, and there were scholars who were executed for giving it to foreign countries. Even then, picking up uranium ore was nothing special. The source of energy is information. (Translated by the authors)
>
> Shinichi Hoshi, "The Net of Voices" (Kadokawa Bunko)

In other words, even if raw material exists, without the knowledge or information on how to extract energy from it, it is just a mere substance. This is why it is argued that information is a worldview. Whether you see the stone in front of you as just an ore, or as a valuable energy source, that difference is information.

What this example shows is that there are things that cannot be seen just by looking at the world from the perspective of matter and energy. Einstein's famous equation $E = mc^2$ describes the relationship between the mass of matter $m$ and energy $E$. This equation explains the energy produced during nuclear fission and fusion, but it alone can neither build an atomic bomb nor unfortunately a fusion reactor. For this purpose, you need information on how to build an atomic bomb and a fusion reactor.

The general term "information" is used in a variety of contexts: Since around 1980 with the spread of personal computers, and since around 1990 with the spread of the Internet, the term "informationization" has become more common. This refers to the use of information technology when companies think about new products or services, or when they want to make individual processes in their operations more efficient and smarter. On a personal level, using information technology as a tool to make daily life more comfortable and effective is also included in informationization.

Unfortunately, there are still not many examples in Japan or around the world of using information technology to revamp traditional workflows or create new business models. Currently, attention is focused on GAFA (Google, Amazon, Facebook, Apple), which have become huge companies through information technology, but the world's information technology engineers intuitively understand that the real transformation brought about by information does not stop at the level of GAFA. They are convinced that information is just on the cusp of further fundamental changes to the world and society we live in.

Looking towards such a future, in recent years, the concept known as "Digital Transformation", or "DX" for short, has been proposed. It aims to launch innovative new businesses using information technology and/or digital technology and fundamentally improve people's attitudes and behaviors towards life. DX is about understanding the true value of information and practicing the worldview of information.

## 1.7 What Computational Thinking is Not

The discipline centered on the worldview of information is "Information Science." Computational thinking is based on the thinking methods created by researchers and engineers of information science who study information. Let's look at how this information science differs from existing disciplines such as physics and mathematics.

### *1.7.1 Differences from Physics*

Let's continue with the example of uranium from the aforementioned book "Voice of the Net" by Shinichi Hoshi. From a physical perspective, the atomic number of uranium is 92, and its atomic weight is 238. The atomic number is the number of protons in an atom, and the atomic weight is the weight of the atom, which is roughly the sum of the number of protons and neutrons. Therefore, we know that uranium 238 is made up of 92 protons and 146 neutrons. As you may know, uranium is a radioactive substance that spontaneously undergoes nuclear fission to split into smaller atoms. This fission occurs slowly in nature, such as inside the Earth.

If we go back to a more fundamental level of matter, matter is made up of atoms, and those atoms are composed of a nucleus made up of protons and neutrons, and a cloud of electrons around it. It is known that atoms generally have an equal number of protons, which carry a positive charge, and electrons, which carry a negative charge, so atoms are electrically neutral. How matter changes, and how energy transforms through heat and motion, is the worldview of "physics", and the knowledge of physics.

Therefore, physics also deals with a lot of "hows", that is, information. But this information is generally descriptive information about the state and movement of matter and energy. This is why physics is said to be descriptive. It does not include information on how to manipulate matter and energy, or how to build machines like nuclear reactors that manipulate and control matter and energy. This kind of information and knowledge for creating something by human hand is dealt with in the field of "engineering", not physics.

The study of observing the state and movement of objects and discovering principles and laws therein is generally called "science." The definition of science is difficult, but let's consider it a method of objectively organizing and understanding things. Things exist first, and then humans understand them. Understanding the laws of nature is natural science, and physics, which deals with matter and energy in the natural world, is a branch of science. Nevertheless, physics is at the core of natural science. In fact, the methodology of natural science has developed based on physics.

Both science and engineering revolve around the loop of hypothesis formation and its verification. In science, to achieve their goals, hypothesis formation involves observing phenomena and inferring the underlying principles or laws. These inferred principles or laws are called "hypotheses." These hypotheses are then tested through experiments to verify their correctness. By going through this loop, principles or laws that explain phenomena more accurately and consistently are gradually being developed.

The situation in engineering is pretty much the same. In engineering, the hypothesis formation involves devising methods to operate the target as intended or to realize the desired function. The correctness of the hypothesis is verified by implementing the method. By going through this loop, the methods to operate the target as intended or to realize the desired function are gradually being developed.

Now, let's shift our focus from matter and energy to information. Information science is a discipline that centers on the worldview of information. In information science, we deal with methods to generate new information based on various information. This is called "inference" or "prediction." Let's introduce a simple example of inference. Let's consider three propositions A, B, and C as follows:

A: Socrates is human,

B: Nightingale is human,

C: If you are human, you get hungry.

From A and C, we can conclude that Socrates gets hungry. From B and C, we can conclude that Nightingale gets hungry. This type of inference is called "deductive reasoning." Also, if we interpret the result of the inference as something that will happen in the future, it becomes a prediction that Socrates will get hungry.

Let's introduce another simple example of inference. Again, let's consider three propositions D, E, and F as follows:

D: Socrates and Nightingale are both human,

E: Socrates gets hungry,

F: Nightingale gets hungry.

From these three propositions, we can infer that "if you are human, you get hungry." Of course, human beings are not just Socrates and Nightingale, but if Socrates and Nightingale get hungry, we might assume that other human beings will also get hungry. This type of inference is called "inductive reasoning." What should be noted here is that the conclusion "if you are human, you get hungry" is not necessarily 100% correct. That's what inductive reasoning is. It only says that it will probably be correct in most cases. If you think of it as a prediction, it's natural that it can be right or wrong.

Some readers may be wondering by now, did not information originally exist in the natural world? Does the concept of information only make sense when there are human beings? Information has two aspects: information as inherently physical existence from the beginning and information as a concept that has meaning only when human beings exist. In fact, there are concepts shared between physics and information science. One of them is "entropy." Entropy, simply put, is the "degree of disorder." For example, a snowflake is a neatly arranged set of water molecules, a state of low entropy. When heat (energy) is added, the entropy increases (in everyday terms, it "melts") and becomes water. Water is a more disordered state than ice. Whether this is disordered or not can be determined by how many possible states that can be taken. Even though there are various types of snowflakes, the number of states called crystals is far fewer than that of states called water.

If left alone, snow will turn into water, but the reverse does not happen. In other words, matter changes from a low entropy (ordered) state to a high entropy (disordered) state, but something needs to be done to cause the reverse. There is a saying, "Spilt water will not return to the tray", which means that water spilt off from a tray will not return to a tray if left alone. In this way, entropy can only change in an increasing direction if left alone. This is called the "law of increasing entropy."

From here, I will make a bit of a bombshell statement regarding the number of states. It is human beings who determine the number of states for molecules contained in crystals and water. In fact, comparing the numbers of states of snowflakes and water are physically meaningless. Under the condition that the number of snowflake molecules and that of water molecules are the same, the numbers of their states are just the same. Moreover, this is because it is impossible to distinguish each molecule in the first place. However, we human beings can recognize snowflakes as having specific patterns, right? Even for water, if you put water in a cup, water should maintain a specific shape. At that time, you human beings likely recognize it as one "state" even though there will be a huge number of states in a physical sense.

Let's explain entropy by using a shuffle of cards. Since it's tough to illustrate with 52 playing cards due to space constraints, consider we have a reduced sized set of playing cards: spades, clubs, hearts, and diamonds, each with a jack, queen, king, and ace. There are 16 playing cards in total. If these cards are separated by type and arranged in ascending order as follows:

(1) ♣J ♣Q ♣K ♣A ♢J ♢Q ♢K ♢A ♡J ♡Q ♡K ♡A ♠J ♠Q ♠K ♠A,

We can say that the entropy is low. If you shuffle this, the order may be changed and the order becomes more random, and the entropy increases.

Now, if separated by number and arranged for the four types in order as follows:

(2) ♣J ♢J ♡J ♠J ♣Q ♢Q ♡Q ♠Q ♣K ♢K ♡K ♠K ♣A ♢A ♡A ♠A,

This can also be said to be a state with low entropy.

What if we would decide that the following is a regular state.

(3) ♡K ♣A ♡J ♠J ♢Q ♡Q ♠A ♢K ♣J ♢A ♠Q ♣K ♡A ♢J ♠K ♣Q?

There are no simple rules. However, physically, there is no reason to distinguish (1) and (2) from each other. Why do we think (1) and (2) are different from (3), although (1), (2) and (3) are certain states of the playing cards, respectively? It's our cognitive mechanism that is creating the "difference." (1) and (2) can be described and recognized by simpler rules, but (3) cannot.

When you shuffle a well-ordered deck of 52 playing cards, it gradually becomes random. The probability of accidentally returning to the original state is very low, practically zero.[5] However, the probability of reaching any specific state is equal and also practically zero.

There are two kinds of things in the world; one is the things that are orderly and understandable to human beings, and the other is the things that appear random. Therefore, "entropy cannot be defined without humans." This phrase was written in "The Order of time" by theoretical physicist Carlo Rovelli. This issue is interesting in itself, but if we delve deeper, we will go off on a tangent, so I will stop here. If you are interested, please refer to the book by Carlo Rovelli.[6]

### *1.7.2 Differences from Science and Engineering*

There is a term "computer science." Can the study of computers be a science? Computers do not exist in nature. They are apparently human-made products of engineering. However, there are already fields called "computer science" or "computer engineering" in the world. Don't you readers feel a slight discomfort with the term "computer science"? I would like to solve this mystery for a moment.

In Japanese, there are expressions like "science and engineering" or "science and technology." In English, there are words like "science and art" as well. Interestingly, when we say art in Japanese, we think of fine arts such as painting, but art in English means technology and skill. Martial arts are techniques of fighting, and liberal arts are techniques (or knowledge) for being liberal. Organizing these terms in the following figure might make it click.

**Column 4 The Law of Increasing Entropy and the Arrow of Time**

Physical equations (such as Newton's equations of motion or the equations of relativity) contain a variable. If this is replaced with, in other words, if time is reversed, the equation still holds. This means that if you rewind a video capturing a physical phenomenon, you cannot distinguish it from playing it forward. However, there is one difference. friction. Without friction, you wouldn't be able to tell if a billiard ball's movement was being rewound, but with friction, the ball gradually slows down and eventually stops. If you rewind this, the stopped ball starts moving, which is strange.

---

[5] The exact probability can be calculated using permutation as 1/P(16,16), which is 1/20922789888000.

[6] Carlo Rovelli: The Order of Time, Penguin Books (2018).

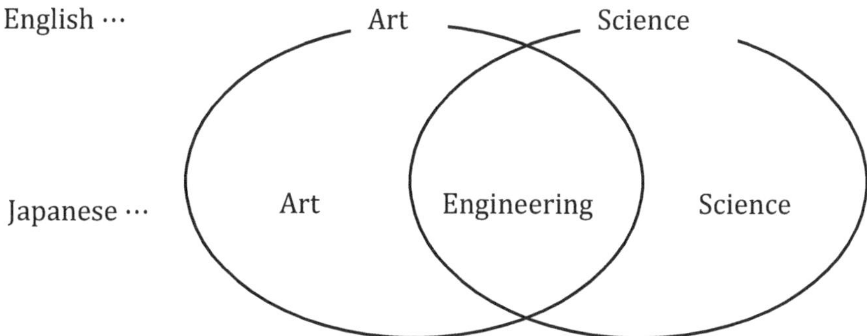


**Fig. 1.9** The relationship between Japanese Art, Engineering and Science, and English Art and Science

Friction is an example of the law of increasing entropy, which means that the ball gradually turns into heat as it moves. The law of increasing entropy is the only thing in physics that determines the direction of time. The existence of time has long troubled physicists. Many books have been published on the "arrow of time".

Recently, a physicist has emerged who claims that this entropy is due to human cognitive function. His name is Carlo Rovelli. This is a unique perspective. He argues that the flow of time for humans exists because humans exist in this universe. The conclusion may still be far off, but I think this is a rare example of a physicist thinking computationally.

Now, I mentioned that physics is called "descriptive", and contrastingly informatics is called "operational", and specifically creates new information, it is called "constructive." Compared to physics, the worldview of information is well expressed by these words "operational" and "constructive."

Computational thinking as presented in this book is part of engineering thinking. In this book, we use the terms "computer science" and "computer scientist." According to Fig. 1.9, we Japanese should instead use "computer engineering."

### *1.7.3 Differences from Mathematics and Logic*

Mathematics manipulates information differently from physics. Mathematics combines one piece of information with another to create new information. There is a divide among researchers as to whether mathematics is something created by humans or something that exists naturally.

There are those who say that every natural number such as 1, 2, 3 etc. is invented by humans, while there are those who believe that they exist in nature even without humans. Logic and mathematics are slightly different from natural sciences, but it wouldn't be unnatural to call them science collectively.

As an example of such mathematics, we will take up Euclidean geometry. Euclidean geometry is roughly the same as the geometry that we Japanese learn in elementary school. To explain Euclidean geometry precisely, it starts with the definition of a point and the definition of a line. A definition is a sentence that precisely determines what properties a concept has and how it is made. Then, properties called axioms are given using that definition. An axiom is the starting point that can be assumed to naturally hold without proof, such as the transitive law and reflexive law in logic.

Furthermore, there are properties called postulates, which are what would happen if we assumed they were correct (these are also sometimes called axioms); for example, "parallel lines do not intersect" and "there is only one line connecting two points." These postulates cannot be proven correct.

Anyone can define anything freely and give axioms and postulates freely. Then, properties derived from axioms and postulates by deductive reasoning are called theorems; for example, the sum of the interior angles of a triangle is 180 degrees, Pythagoras' theorem, etc. However, if contradictory axioms are given, contradictory conclusions will be derived, and if unrelated postulates are given, no meaningful conclusion will be derived. These meaningless conclusions cannot be called theorems.

As another example, let's look at the definition of natural numbers. There are various ways to do this, but here we will introduce a definition using sets. Since they are natural numbers, we create them in the order of 0, 1, 2, etc. We use existing information to create new information. First, let's assign the symbol '0' (zero) to the empty set. Let's write it as $0 = \{\}$. The empty set does not contain elements and is unique, that is, there is only one empty set in the world.

Since all numbers can be represented as sets, let's define $n + 1$ as $n \cup \{n\}$. This term means the set that is the union of the set that represents $n$ and the set that has only $n$ as an element. In this way, the set of $n + 1$ has one more element than $n$, and the newly added element $\{n\}$ is obviously not included in $n$ . Therefore, when you have made $0, 1, \ldots, n$ in order, you can always create a set that is different from the sets you have created so far. Let's see what kind of sets are actually being derived in order.

$$\begin{aligned}
1 &= 0 \cup \{0\} = \{\} \cup \{\{\}\} = \{\{\}\} \\
2 &= 1 \cup \{1\} = \{\{\}\} \cup \{\{\}\} = \{\{\}, \{\{\}\}\} \\
3 &= 2 \cup \{2\} = \{\{\}, \{\{\}\}\} \cup \{\{\{\}, \{\{\}\}\}\} = \{\{\}, \{\{\}\}, \{\{\}, \{\{\}\}\}\}
\end{aligned}$$

It goes like this. However, there are too many brackets and it's hard to understand what's going on. Would it be easier to understand if we rewrite it as follows?

$$\begin{aligned}
1 &= 0 \cup \{0\} = \{\} \cup \{0\} = \{0\} \\
2 &= 1 \cup \{1\} = \{0\} \cup \{1\} = \{0, 1\}
\end{aligned}$$

$$3 = 2 \cup \{2\} = \{0, 1\} \cup \{2\} = \{0, 1, 2\}$$

In this way, it should be clear that the newly created set can be written as follows:

$$n + 1 = \{0, 1, \ldots, n\}.$$

It is also apparent that $n + 1$ is different from any of $0, 1, \ldots, n$. Here, $0 = \{\}$ and $n + 1 = n \cup \{n\}$ correspond to the axioms, and

$$\begin{aligned} 1 &= \{0\} \\ 2 &= \{0, 1\} \\ 3 &= \{0, 1, 2\} \end{aligned}$$

correspond to the theorems.

In mathematics and logic, new knowledge and information called theorems are created from axioms and postulates. Above, I wrote that creating new information based on specific procedures is called computation, but since theorems are new information, the operation of creating theorems can also be considered computation. In this sense, part of mathematics and logic can be considered the equivalent of computational thinking or the worldview of information.

So, what are the points where mathematics and logic differ from computational thinking? In mathematics and logic, there are cases where "you don't have to know the specific procedure of computation." If you are good at mathematics, you may have encountered cases where you can think of many ways to solve a problem. That's right, there are generally multiple ways from axioms and postulates to theorems. Therefore, "a method based on a specific procedure" usually means that you know at least one of the several specific procedures. However, there are also cases where you only know that there exists a path consisting of specific procedures, but you don't know what it is.

Mathematicians sometimes use the phrase "principally possible." What it means to be computationally possible in principle is that it doesn't matter how much time it takes, or how much memory or disk space is consumed. There is a big difference between what can be computed in principle and what can be computed in reality. Even if it is computationally possible in principle, if the computation time takes hundreds of years, or if the required memory size is larger than the number of atoms in the universe, the execution of the computation is practically impossible. In computational thinking, we are aware of whether the computation can be completed within a realistically acceptable time range, or whether the computation can be executed with the memory capacity and disk capacity that are realistically available. Memory capacity and disk capacity correspond to the maximum amount of data that must be recorded during the computation. Mathematics covers ideal and limit states, but computational thinking is limited to the range that can be calculated in practice.

In logic and mathematics, there is a proof method without calculating a certain value known as "reductio ad absurdum." In reductio ad absurdum, we take the approach of starting the proof without touching the specific computation procedure, just by assuming the existence of a certain mathematical concept. For example, when trying to prove a certain proposition to be correct, if it is difficult to prove it directly, we consider what conclusion would be under the assumption that the proposition is wrong. Then, if a contradiction is derived by assuming that a certain proposition is wrong, it is concluded that the proposition was correct. Reductio ad absurdum is more straightforwardly called proof by contradiction. Here, a contradiction refers to the state that for a proposition P, "P" and "not P" are held at the same time.

**Reductio ad absurdum (proof by contradiction)**

To prove a proposition P, instead of directly showing its correctness, it is a method of proving that P is true indirectly by showing that it contradicts when assuming "not P".

Now, let's say we want to prove the proposition "There are infinitely many prime numbers." However, even using the fastest supercomputer in the universe, you cannot exhaustively generate an infinite number of primes. It takes an infinite amount of time to generate an infinite number of primes. So, it's hard to prove directly. Therefore, we assume that "there are only a finite number of primes, and all of these finite primes are $p_1, p_2, \ldots, p_{n.}$" That is, the negation of the proposition to be proved has been assumed. The point here is that it does not provide any specific method or value for calculating $p_1, p_2, \ldots, p_n$, it only assumes that such finite primes exist. Next, consider the new number $p = p_1 \times p_2 \times \cdots \times p_n + 1$. Then, for all $i = 1, \ldots, n$, it is clear that $p$ is larger than $p_i$, and $p$ cannot be divided by $p_i$ because it always leaves a remainder of 1. The new number $p$ actually meets the conditions of a prime number. Therefore, it contradicts the assumption that "there are only a finite number of primes, and the entire finite number of primes are $p_1, p_2, \ldots, p_n$." Therefore, it was concluded that "there are an infinite number of primes." This is a famous proof by Euclid for the proposition that "there are an infinite number of primes."

Not only reductio ad absurdum is often used in mathematics, but also it appears in detective novels. For example, suppose there is information that only the culprit would know in a murder case and the murder weapon has not been announced. If the suspect is assumed to be "not the culprit", the suspect should not know about the weapon. Let's say the detective cornered the suspect and made him/her say "I couldn't possibly have a gun". The weapon could have been a knife, and the only one who could specify a gun would be the culprit, so such reasoning becomes valid. In other words, if the suspect is assumed to be "not the culprit", it contradicts the fact that the suspect cannot know that the weapon is a gun. Therefore, it is reasoned that the suspect must be the culprit.

In mathematics, computation was a powerful method to generate new information, but logic, represented by reductio ad absurdum, is also a powerful method to derive new conclusions and generate new information. Computation and logic, both powerful methods to generate new information, differ greatly in the point of whether computation calculates specific values in order according to specific procedures and reaches a conclusion. Then, the message conveyed in the book is that thinking and executing these specific procedures would be very useful in everyday life.

### *1.7.4 Difference from Philosophy*

Philosophers often talk about things "in principle." It's similar to mathematics.

We are going to discuss the thought experiment of "Chinese Room" by John Searle,[7] a typical argument of philosophers. This is a fictional experiment to argue that machines never "understand." The setting of the Chinese Room is as follows.

- There is an American person in the room who does not understand Chinese.
- A piece of paper with a question written in Chinese is thrown into the room.
- The person does not understand Chinese but has a thick English manual written in English about Q&A in Chinese.
- The person follows the Chinese characters on the paper with a question in order (it's hard to find unreadable characters, but let's ignore that), works intensively according to the instructions in the manual, and eventually gets a response in Chinese.
- The person writes down the response in Chinese on a piece of paper and returns it out of the room.

From the perspective of a person outside the Chinese Room, if you put in a question in Chinese, you get a response in Chinese. So, the room seems to understand Chinese. However, when you peek into the room, there is the person inside who does not understand Chinese at all, and of course, does not understand the content of the question or answer. Searle argues that AI programs are the same as the person inside, even if they appear to behave intelligently, they do not have intelligence.

There are various counterarguments from the AI side, one of which is about computational complexity and descriptive quantities. I don't think a manual that anticipates all questions can be created in the first place, but even if it could, how thick would it be? One AI researcher wrote a paper trying to estimate that amount and proved that even simple addition exceeds the total number of atoms in the universe, as it is difficult to measure the description of a general manual.

Searle does not have an estimate of the amount of work. He describes the thought experiment "in principle." The problem is discussed as if an answer can be created in a few minutes or at most a few hours, but it turns out that this is not the case.

In other words, the difference between philosophy and computational thinking is whether to consider the resources (memory and computation time) required for computation. Estimating this computational quantity is quite important for us to solve various problems in our daily lives.

---

[7] American philosopher (1932–2025).

The "Frame Problem" is also famous as a philosophical problem discovered by AI researchers, John McCarthy and Patrick Hayes. Originally, it was a problem that it is difficult to describe how the situation is changed by performing a certain action. Despite only a part of it being changed, a large amount of unchanged framework (frame) must be described, hence it was named as such. In other words, the Frame Problem means that when trying to describe an action, the effort to explicitly describe both what is changing and what is not changing, and to distinguish between them, may increase exponentially in description and inference.

If we don't think about specific computations, we don't feel the frame problem in our daily lives. That was why philosophers didn't notice it. However, when McCarthy tried to write the specific computation down as an AI program, he faced this problem. Even in the simple problem called "block world", where they were just rearranging blocks in early AI research, they couldn't write an AI program that realized specific computation rules.

One of the important things in computational thinking is to know that there are limits to computation. Even if the problem is describable in principle, there will be no end to the list of specific conditions. This may not have been a problem in mathematics or philosophy where it is finished with "it is possible in principle." This problem cannot be given a solution even by computational thinking, but it is important to recognize that there is such a hard problem. When talking with friends or family in daily life, you may find your understanding of "it is possible in principle" is at odds with theirs and you may not even realize it.

### *1.7.5 Differences from Programming*

It seems that there are many people in the world who understand that computational thinking is the ability to write computer programs. But that's a mistake. There are many people who can program but cannot think computationally. Conversely, there are many people who can think computationally but don't need to write a program. So, what is the difference between being able to program and doing computational thinking?

First, a "program" is a blueprint for realizing the computation you want to achieve, which is a procedure that describes the specific steps in order. "Programming" refers to the task of creating a program. The machine that executes the program is called a "computer." The languages (natural languages) used for human communication include English, Arabic, and Japanese. Similarly, writing programs, "programming languages" such as Python, Java, and C are used. In these programming languages, like mathematics, a very small number of well-defined elements are given at first, and from there, elements are combined according to strictly defined rules to gradually construct complex computations. Strict rules for executing a program are also set in detail, and new information is generated one after another as the execution proceeds. This is what it means to compute in real life.

Assuming we have a Python program here, a program that computes and generates the same information as the Python program could be written in Java or C. For example, a program that makes an LED blink at a frequency of once per second can be written not only in Python, but also in Java, or C. This is the same as being able to greet or express gratitude in English, Arabic, or Japanese. The reason why programs having the same function can be written in different languages is because in the brain of a programmer, there is a procedure at the abstract level computation that does not depend on the existing programming languages used including Python, Java, and C. Such a procedure at the abstract level computation is called an "algorithm."

As an example of an algorithm, let's introduce Euclid's algorithm, which is also known as the oldest one in the world. This is to find the greatest common divisor (GCD) of two natural numbers (positive integers) as follows.

> For two natural numbers $a, b(a \geq b)$, if the remainder of $a$ by $b$ is $r$, the greatest common divisor of $a$ and $b$ is equal to the greatest common divisor of $b$ and $r$. Using this property, if the calculated value of the remainder of $a$ by $b$ is $x$, then $b$ is set to the new $a$ and the remainder $x$ is set to the new $b$. Such that, the calculation of the reminder of $a$ by $b$ is repeated until the reminder becomes 0. Then, $b$ when the remainder becomes 0 is the greatest common divisor of original $a$ and $b$.

This is probably one of the things that everyone has learned in school. What I want you to pay attention to here is the method for explanation of the computation procedure. In the explanation, there is no mention of specific programming languages such as Python, Java, and C. However, based on the explanation, everyone can implement Euclid's algorithm in Python, Java, or C. This is because the computation procedure is described at an abstract level.

The process of thinking about an algorithm and implementing the algorithm in a specific programming language is called "programming." The first half, where you think about the algorithm, is sometimes called "design" or "planning", and the second half, where you implement it in a specific programming language, is sometimes called "programming" or "coding." In the latter part, whether you are programming alone or with multiple people, various programming methodologies and practices have been developed. Those methodologies and practices lead to the field of study known as software engineering.

According to Wing, "thinking like a computer scientist" is the definition of computational thinking. In other words, it is the application of the thinking methods developed by researchers and engineers in informatics, who study information, to practical tasks in real daily life. It is the computer scientists who have come up with many efficient and readable algorithms and have devised programming techniques or methods. Furthermore, computer scientists have not only pursued programs, but also the hardware that runs the programs.

The machine called a computer that executes a program is a physical entity wired with electronic components made of semiconductors like silicon and metals like copper or aluminum. Since a computer can only operate within such physical constraints, it must consider hardware, fast and power-saving circuits, technology to pack as many components as possible into a small space, easy-to-control configurations (good compatibility with software), and suitability for mass production.

Therefore, "thinking like a computer scientist" includes this hardware part, and that's why the thinking method of computer scientists can be applied to our daily life surrounded by physical constraints, and that's where the value of computational thinking lies. Computational thinking is not just about programming.

Of course, there are commonalities between computational thinking and programming. To create a good program, you have to think about various things. Efficiency (fast operation, less memory required) and readability (easy to understand, easy to maintain) are important, and in particular, efficiency is closely related to computational thinking.

What you need to be careful about when programming is the phenomenon of exponential explosion mentioned in Sect. 1.3. When you are writing a program, have you ever thought that you should let the computer do simple repetitive tasks because that's what computers are good at? I think it's a common occurrence in programming. This whisper of the heart is basically correct, but if you let the computer do simple repetitive tasks without thinking deeply, you will fall into the phenomenon of exponential explosion, and no matter how fast the computer is, it will take so long to finish the computation that you will lose your motivation. Thus, when writing a program that lets the computer do simple repetitive tasks, think carefully to avoid exponential explosion. Also in daily life, if you are aware of actions or situations that could lead to an exponential explosion, you have to avoid inefficiency and irrationality.

**Column 5 "Computer" Once Meant a Person Who Computes**

The word "computer" originally referred not only to machines but also to people who performed calculations. Renaissance astronomers such as Johannes Kepler relied heavily on numerical work to calculate planetary positions. At that time, assistants known as "computers," meaning people who carried out calculations, helped perform these tasks.

Until the advent of commercial electronic computers, the term "computer" referred to people who performed mathematical calculations. From the late nineteenth century, particularly in the United States, human computers worked on calculations for astronomy, meteorology, and ballistics. They organized into teams ranging from a few members to several hundred, dividing tasks to carry out large and complex calculations in parallel. During World War II, as many male researchers were drafted, the occupation of "computer" became predominantly female. In Japan, female students also played key roles in computing altitude and azimuth tables used for aerial navigation of the Ginga bomber.

These computers used mechanical and electromechanical calculators, and later early electronic computers, to support their work. As fully electronic computers became available, many of these human computers transitioned into the emerging profession of computer programmer. Notably, the first six programmers of ENIAC, the Electronic Numerical Integrator and Computer, were women who had been recruited from ballistic computing teams.

Since the late twentieth century, individuals with extraordinary mental-arithmetic skills have sometimes been called human computers. Over time, the meanings of compute, computation, and computer have evolved.

This evolution, from computer as a human occupation to computer as a machine, reflects how advances in technology have continually reshaped both language and society.

**Column 6 Women's Pioneering Role in NASA's Computing History**

*The Calculating Stars*, a celebrated science fiction novel, received the Hugo, Nebula, and Locus Awards. Set in an alternate history where a massive meteorite strikes Earth and triggers a space colonization program, the novel accelerates the space program's timeline. It follows a woman working as a 'computer girl' in NASA's computing division, tracing her remarkable journey from performing orbital calculations to becoming an astronaut who travels to the Moon. In this context, 'computer girls' refers to women who carried out critical trajectory and orbital computations. The story unfolds in the 1950s, before the real U.S. space program, depicting a time when humans, not machines, were the primary 'computers.'

The novel's themes resonate strongly with those in the film *Hidden Figures*, which is based on a true story. This film recounts the challenges and triumphs of three exceptional African American women at NASA during the segregation era. As human computers, they played vital roles in the space race against the Soviet Union, producing groundbreaking calculations. When NASA began adopting electronic computers, one of them mastered FORTRAN and taught programming to her colleagues. She eventually led a team of thirty women, including herself, in the new programming division and was promoted to its supervisor. On a crucial launch day, when astronauts doubted the machine's calculations, her precise manual calculations confirmed the accuracy of the mission's trajectory. In 2016, NASA honored her by naming the Katherine Johnson Computational Research Facility at the Langley Research Center, commemorating a true pioneer in both space exploration and computing.

## 1.8 Deep Learning and Generative AI

One of the reasons for the recent attention to AI technology, which has had a great impact on society, was the announcement of ChatGPT 3.5 by OpenAI in November 2022. Since the Japanese edition of this book was published in January 2022, it did not include descriptions of ChatGPT, generative AI, and large-scale language models (LLM). We would like to discuss this topic in the English edition of the book.

Computational thinking in this book means thinking like a computer scientist or applying methodologies for generating and analyzing information by computer to everyday life in general. Efficiency at the symbolic processing level or information processing literacy is at its root. In contrast, deep learning is the process of collecting a large amount of training data or examples to train a huge neural network. In the Transformer framework for generative AI, optimization of the functions with hyper-multi-dimensional vectors as input and output is the underlying principle.

Thus, whereas traditional information processing is dominated by methodologies in which humans write down their knowledge and experience as rules and algorithms in order to achieve the desired function, deep learning and generative AI are now able to achieve the same or more complex functions through learning with a large amount of data. In other words, the cost and accuracy of collecting and learning a large amount of data has come to exceed the cost and accuracy of describing knowledge as rules, due to the spread of the Internet, advances in machine learning technologies, and improvements in hardware performance.

How will computational thinking change and evolve in the current and future era of deep learning and generative AI?

First, since deep learning makes inference from some examples, it is impossible in principle to learn and infer 100% correctly. It is known that two types of improper learning can occur (see Column 7). One is called overfitting, in which the generalization ability becomes low due to over-adaptation to the given data (fitting too restricted examples), resulting in misjudgment even though the data is correct. The other is the opposite of overfitting, and is called overgeneralization, in which the system generalizes too much from the given data (generalizes beyond the range of correct answers), and even incorrect data are accepted as correct. In reality, learning is performed with a mixture of overfitting and overgeneralization.

By intentionally creating data that falls within the small difference between the range that a neural network that has completed training judges to be correct and the true correct answer, it is possible to deceive the network. In fact, many examples have been reported of slightly modifying an image of a school bus to make it misjudged as a camel or adding noise to a road sign to make it misidentified as a different sign.

In addition, neural networks are subject to the biases of the training data used for learning. For example, there is gender bias in that pilots are male, and cabin attendants are female, or medical doctors are male. There are also statements that are statistically correct but should not be expressed humanely, including racial bias in certain professions. In order to deal with such principles related to ethics, morality, justice, and religion without offending public order and morality, it is necessary to properly treat meta-level evaluations and reasoning such as values and ethics. It would be difficult just to describe and learn them as a language, but there are many implicit parts that are difficult to describe as a language in the first place, and it would be extremely difficult to capture them in machine learning.

This is why it is necessary to collect a large amount of good-quality training data and examples. However, at the data collection stage, it is not clear how the data will be used and processed for learning and inference. For a computer, it is sufficient to collect as much data as is relevant to its purpose, but this is not obvious beforehand. Therefore, it would be better to collect all the good data in the world in advance, but this is not possible. This can be called the frame problem in the era of deep learning and generative AI, which has not been solved so far.

Since human beings share embodiment and living environment and have learned various constraints from years of living together, this real-world experience and learning provide the criteria for deciding what range of data to consider, and the frame problem is rarely apparent (but not completely gone). Since generative AIs do not live their social lives as living people, it is difficult for them to share experiences and learnings as humans do. For example, learning about love from a romance novel is different from actually experiencing love, and understanding a dish by reading a cooking recipe is different from actually cooking it and tasting it. The same applies to the appreciation of music and paintings.

Therefore, in front of a neural network, a computer scientist who is trying to achieve the desired deep learning should pay attention to the following in order to make the neural network learn and infer as correctly as possible.

What kind of data do you want the neural network to output or infer?

What kind of data and conditions should be input for that purpose?

How and where should such data be collected?

What kind of neural network configuration should we construct?

What kind of data should be given to the neural network, and how?

Until now, computational thinking has focused on efficient handling of computational resources. Computational thinking in the era of deep learning and generative AI requires us to be aware of how to correctly handle the meaning of data. We believe that such new computational thinking will become important in the way we deal with deep learning and generative AI.

In symbolic processing, which has been the premise of computational thinking up to now, the issue of data representation (i.e., what is a symbol and what is a feature) was decided by humans in advance, so the inference process and the meaning of symbols are understandable to humans. On the other hand, deep learning can automatically acquire data representations of an embedding space consisting of super-multi-dimensional feature vectors. Therefore, while it is possible to acquire tacit knowledge that is difficult to verbalize, the inference process and the meaning of feature vectors are difficult for humans to understand and become a black box. The object of computation in traditional computer science is a symbol, and the hyper-multi-dimensional feature vectors that are the object of computation in deep learning may correspond to signals. Because of signals, to some extent, we must accept that the contents of a neural network are a black box.

Novel computational thinking may be to understand not only the strengths of deep learning and generative AI, but also its weaknesses and shortcomings, and to cope with them.

**Column 7 Overfitting and Overgeneralization**

Consider learning to recognize images of cats. By presenting various images, the goal is to have the system identify only the images of cats as "cats." Suppose the target range of correct answers is within the circular range shown by horizontal lines in Fig. 1.10. Since the system cannot know this correct range, it must infer (learn) from the given examples. But precise learning is difficult when relying solely on positive examples. For instance, although the provided examples (indicated by the small squares in Fig. 1.11) are covered, they might be too close to the examples, resulting in low generalization (this is called overfitting). In overfitting, even correct examples might be marked incorrect. Conversely, aiming too much for generalization may exceed the range of correct answers, leading to incorrect data being judged as correct (Fig. 1.12). This is called overgeneralization.

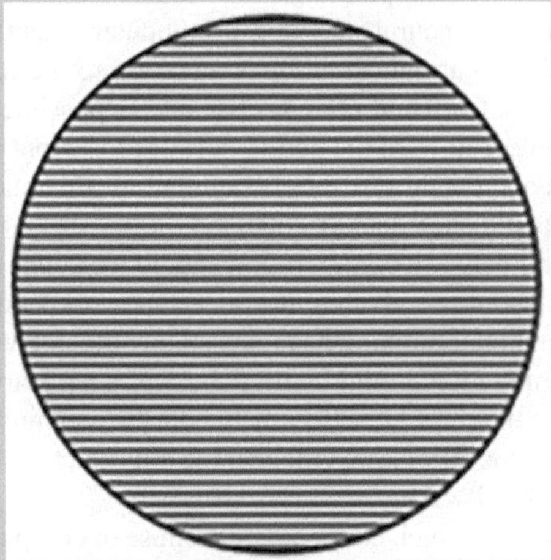

**Fig. 1.10** Target: Any point within the circle is the correct instance of the category which is the target of learning. However, the correct target is not known to the system

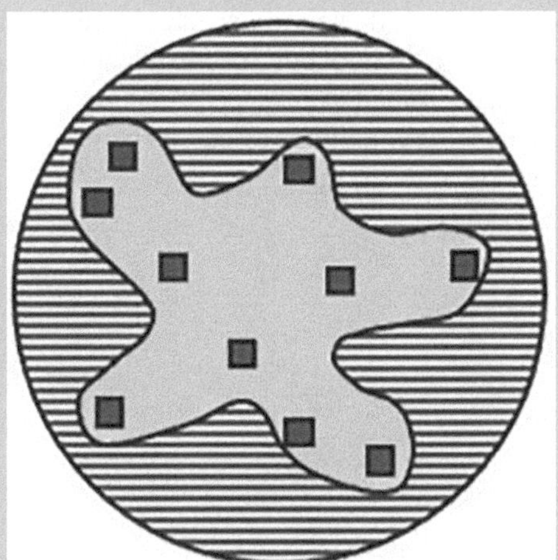

**Fig. 1.11** Overfitting: Small squares represent correct instances given to the system. If the system tries to be precise only for those instances, the result of learning is too narrowly focused on the target area

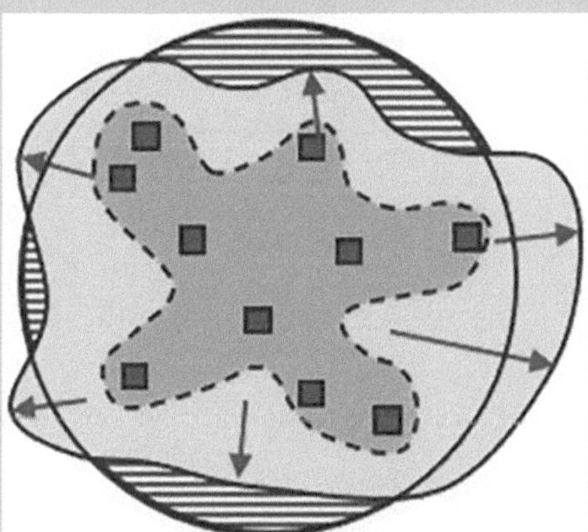

**Fig. 1.12** Overgeneralization: When the system tries to generalize from the given instances too much, it leads to overgeneralization, which should be avoided

To deceive the results of deep learning, one can create examples that differ from the correct answers. Especially in areas where there is a lack of generalization (indicated by the horizontal line in Fig. 1.11), even though it is an image of a cat, it might be judged as something else. On the other hand, if one provides examples where overgeneralization has occurred (areas exceeding the correct range in Fig. 1.12), it might answer "cat" even though it is an image of something else.

There are many studies conducted to deceive the results of deep learning. Examples include making a "school bus" be misrecognized as an "ostrich", or a "stop sign" being misidentified as a "speed limit sign." These methods are more elaborate than the misrecognitions discussed here and will not be detailed as they exceed the scope of this book.

# Chapter 2
# Practice of Computational Thinking

**Keiji Hirata and Hideyuki Nakashima**

**Abstract** Applying computational thinking in daily life and work means understanding the things and events around readers by using the way computers work as an analogy and engaging with them actively. As a result, readers gain an incredibly clear understanding of these things and events and can compute very efficiently. To achieve this, four essential concepts are indispensable. This chapter introduces three of them: abstraction, modeling, and virtualization (the remaining one will be introduced in the next Chap. 3). Computational thinking encompasses not only arithmetic operations like addition and multiplication, but also the broader process of generating new information from existing information. These three (four) key concepts can be seen as the fundamental operations for creating new information. Some may not be familiar with the terms abstraction, modeling, and virtualization in everyday conversation, but here are examples showing how these concepts function right around us. At the end of this chapter, we introduce the field of information design to reaffirm the characteristics of computational thinking. Readers will see that the goal of information design—to make the representation and communication of information and knowledge more flexible and richer—is closely connected to computational thinking as a methodology for generating new information.

## 2.1 What It Means to Be a Computer Scientist

I would like to explain the thought required to practice computational thinking in everyday life.

Computational thinking means thinking like a computer scientist. However, this does not mean being familiar with computer hardware, software, and systems.

Computer scientists, who have a deep and broad understanding of computers and feel familiar with them, often anthropomorphize the mechanisms and operations of computers. For example, they may say there are relationships called parent processes

K. Hirata (✉) · H. Nakashima
Mirai Share Co., Ltd. and Future University Hakodate, Hakodate, Hokkaido, Japan
e-mail: hirata@miraishare.jp
URL: https://www.miraishare.co.jp/



H. Nakashima and K. Hirata (eds.), *Computational Thinking*,
https://doi.org/10.1007/978-981-95-5962-6_2

and child processes, and a program is 'running'. It is even common to say things like 'killing' a process. Conversely, computer scientists express the state of the real-world using computer terms. Saying, for example, "this recipe should be refactored", and "we want to give this whole business a reboot", etc.

Thinking like a computer scientist includes creating metaphors for real-world things and events using the mechanisms and operations of computers. There is an English proverb,

*If all you have is a hammer, everything looks like a nail.*

This means that if you only have one means, you will try to solve all problems with that one means, which may be inefficient and ineffective. Computer scientists, in a good sense, try to understand everything in the real world by metaphors using the mechanisms and operations of computers.

What I want you to note here is the difference between a hammer and a computer. In common sense, a hammer only has the function of hitting a nail, but a computer is versatile. In principle, a computer can perform any calculation and realize any function. Therefore, computer scientists, who have a deep understanding of the mechanisms and operations of computers, can create a metaphor for any real-world things and events using the appropriate mechanisms and operations of computers without compromising their essential properties and relationships.

## 2.2 Abstraction

### *2.2.1 Meaning of Abstraction*

What does it mean to create a metaphor for something using the mechanisms and operations of a computer? Here are three concepts that become important. They are abstraction, modeling, and virtualization, all of which involve applying something to the original thing and creating something new.

By properly abstracting, modeling, and virtualizing real-world things and events, you can correctly understand and imitate (simulate) the real world through symbols on a computer. Here, the symbols on a computer refer to variables and numbers that appear in a program, or their arrangements (arrays or structures), but they are always required to correctly correspond to real-world things and events throughout the execution of the program.

To accurately forecast the weather, you need to accurately observe the temperature, humidity, atmospheric pressure, wind speed, etc. in various places on the earth, and throw them into a huge set of differential equations. Prior to this, there is the task of expressing the physical phenomenon of weather as a differential equation with parameters (or intermediate variables). In the real world, there are a wide variety of physical parameters, as well as factors such as sounds heard on the spot (soundscapes), the body temperature of the people living in the relevant area, the number of dogs kept, the proportion of people who like Mozart, etc., but common sense tells us that these pieces of information do not have a significant impact on the weather, so it is safe to assume that there is no need to incorporate them as parameters in the differential equation for weather forecasting. Thus, meteorologists have simply selected those parameters that are most heavily involved in the weather, such as temperature, humidity, atmospheric pressure, and wind speed.

According to the sixth edition of 'Kojien', one popular Japanese dictionary, abstraction is defined as 'a mental action that grasps a certain aspect or property of a thing or representation apart from reality'. In the above example, abstraction refers to the process of selecting parameters that are considered important, by setting a certain perspective, and selecting parameters that are thought to have a large effect on the phenomenon being observed. In other words, abstraction is defined as focusing on important information and ignoring complex details. When creating a differential equation, we simply ignore parameters that are not important or that are unrelated to the phenomenon being observed. This is called omission.

Furthermore, the word abstraction is used in a sense other than selection of parameters. That is, when parameters that are not independent of each other are found, they are combined into one to reduce the number of parameters. For example, consider the course of a marathon race. The distance from the start to the goal is fixed at 42.195 km, but the shape and elevation of the course vary from city to city. As a parameter to represent where the runner is running on the course, a point in a three-dimensional Cartesian coordinate space can be used. In other words, we need three variables, X, Y, and Z, to which values representing the runner's location are assigned. However, the course the runner runs is a single path, and the runner cannot freely exist at any points in the three-dimensional Cartesian coordinate space. That is, parameters X, Y, and Z are not independent of each other, and there are certain constraints on them. Considering these constraints, the runner's position can be uniquely determined by giving only one variable: how far from the start the runner is on a single path. As a result, the runner's position has been abstracted as a single parameter, how many kilometers from the start. Information about the shape and elevation of the course, which varies from city to city, has been discarded.

**Column 8 Definition of "Abstraction" in a Dictionary**

'Super Daijirin', a commonly used Japanese dictionary, defines abstraction as follows: The act of focusing on certain properties, commonalities, or essences of things or representations, and understanding them by extracting them. This process also involves the action of excluding other unnecessary properties (= omission), so abstraction and omission form two aspects of the same action.

Example: "what appears in the meaning or judgment is a part abstracted from the original experience."

Opposite: concrete. See also 'omission'.

### 2.2.2 *Difficulty of Appropriate Abstraction*

Regarding the parameters selected when abstracted in this way, the physical variable of temperature, for example, has a strictly consistent definition and is implemented as a corresponding variable in the program. As long as the calculation is carried out according to the differential equation, the variable corresponding to the temperature always correctly represents the temperature throughout the execution of the program.

To make a more accurate weather forecast, it might be effective to consider the effects of clouds, aerosols, cosmic rays, terrain, buildings, trees, and traffic volume. However, how should we express information about clouds and terrain in differential equations? It does not seem possible to simply represent them with real number variables in the same way as temperature and humidity. Even if we express information about clouds with specific terminology like "cumulonimbus" or "streaky clouds", we cannot incorporate it well into differential equations or forecast calculations. We would probably have to use vectors and matrices and set complex boundary conditions. However, it is generally very difficult to give a physically strictly consistent definition to the parameters representing the clouds and implement them as arrays or structures in the corresponding program.

### 2.2.3 *Good Thing About Abstraction*

So, what is the benefit of the abstraction that computer scientists are proficient in? It is that when you abstract, different things look the same. This ability to see the same is inevitable to create a metaphor for real-world things and events using the mechanisms and operations of computers. For example, if you add 8 GB of memory to 4 GB of memory, the total is 12 GB. If you load 4 kg of rice and 8 kg of water onto a truck, the total is 12 kg. If you buy a $400 ring and an $800 necklace, the total is $1,200. If 8 people get on a bus with 4 passengers, the total is 12 people. All these real-world calculations are achieved by the same process, which involves first abstracting the relevant factors into the numbers 4 and 8, calculating using the addition formula $4 + 8 = 12$, and then returning the value 12 to the real world again (called concretization or instantiation). By selecting only the property of value, that is, setting the perspective of quantity and focusing only on that value, various phenomena in different domains such as memory expansion, total weight, total amount, and total number of passengers appear the same.

## 2.3 Modeling

### *2.3.1 Meaning of Modeling*

By abstracting things and events in the real world, what we get next is a model. We call the process of obtaining a model from the real world 'modeling'. Normally, when we say "model", many people likely think of human models appearing in fashion shows, but, in this book, let us assign to the word model the meaning of a substitute for the real thing with respect to functions and features, which works for understanding and explanation in a broad sense.

Even if we assign such a meaning to the word model, the concept may be still unclear, so let us consider Newtonian physics, one of the most successful models in human history, as an example. Long before Newtonian physics, around 350 BC, Aristotle claimed that all matter in the world was made up of four elements: fire, air, water, and earth. In China, it was thought that the world was made up of fire, water, wood, gold, and earth, according to the theory of five elements. However, these theories could not explain certain mysteries, such as why a small iron ball and a large iron ball and a feather all fall at different speeds, and why the moon does not fall to the ground. Then, in the 1660s, Newton discovered (or "extracted") the concepts of mass and force. Until then, people had tried to understand the world through visible things, but Newton tried to understand the world with essential concepts that are invisible. Newton proposed the famous formula $F = ma$ ($F$ is force, $m$ is mass, and $a$ is acceleration). Furthermore, he also discovered (or "invented") the concept of calculus to make it possible not just to understand the relationship between objects, but also to predict movement. This calculus can be seen as a technology to predict the temporal change of force, mass, and acceleration using a model.

### *2.3.2 Three Types of Models*

The models actually used in the world are broadly classified into three types: mathematical (or representational) model, numerical model, and concrete model. A mathematical model is one that expresses the behavior or state of the target object in mathematical formulas, such as Newtonian physics (classical mechanics) and linear regression used in data analysis.

A numerical model expresses the behavior or state of the target object with an algorithm. Describing by an algorithm means using computational means such as sequential execution, iteration, conditional branching, recursive call, probabilistic state transition, and multi-agency. An earth simulator and cellular automaton are examples of this.

A concrete model is one that substitutes the target object with another physical object. For example, model organisms like guinea pigs; miniature airplanes or rockets used in wind tunnel experiments to test aerodynamic performance; and celestial globes that faithfully reproduce the movements of the sun, moon, and stars visible in the sky (such as Ptolemy's famous celestial globe). The human models that appear in fashion shows can also be considered concrete models, as they demonstrate and simulate what it would be like for someone to wear certain clothes.

The purpose of obtaining and considering models is to understand, explain, and predict the phenomena of the real world from a certain perspective, which is something that humans have an essential desire to do. In Section 1.7.2, we stated that "(the reason computational thinking is important is because) computation creates new information based on specific procedures." Prediction is a type of information generation, and it is an important activity for humans. However, it is generally very difficult to invent a model for generating new information or making predictions. For example, compare the task of judging which of two dishes is more delicious with the task of actually cooking a dish that is more delicious than another. It is immediately clear that the latter task is harder, as it is much more difficult to actually cook food than to judge it. The same is true for earthquakes and economics, as it is well known that it is much more difficult to predict what will happen than to analyze what has happened. Thus, practicing computational thinking and generating new information is not an easy task.

Then, let us think a little more about prediction in mathematical models. First, by abstracting things in the real world, we select parameters, and relationships between parameters, that represent the state and changes we are focusing on. In the case of Newtonian physics, the parameters are force, mass, and acceleration, and the relationship between parameters is $F = ma$. The whole of these parameters and relationships is called a structure. Since the real-world changes over time, or in response to certain action, it is also necessary to consider the operations that change the corresponding structure. In Newtonian physics, computing calculus corresponds to performing the operations. The structure and the operations that alter the structure are combined to form a predictive model. With the use of calculus, you can predict how an object changes over time. The correspondence between the structure and operations included in a model and the elements and changes in the real world is called interpretation. Thus, a mathematical model is also said to be an interpreted mathematical structure.

### *2.3.3 From Modeling to Science*

By improving a model, it can be made to function more appropriately, enabling humans to explain the phenomena of the real world more skillfully. This joy of being able to explain invokes a kind of aesthetic sense, and it is said that this led to the discovery of science in 16th century Europe. To facilitate the joy of being able to explain, the following three methodologies, each containing a "not" have been adopted among scientists.

- Not being concerned with the purpose (though science may originally have been used for agriculture or war, the ability to explain itself became the purpose),
- Not pursuing preciseness (the simplicity of the model was prioritized even if the prediction was slightly inaccurate), and
- Not caring that the conditions set are artificial (e.g., experiments on dropping iron balls in a vacuum, experiments using miniatures, or introduction of mass points).

By developing such methodologies, science has been proposing better models, that is, more concise models that can provide a more thorough explanation. Gradually, as the concept of science was formed, a certain consensus was also formed about the attitude of scientists in pursuing science. Definitions regarding the scientific attitude were given from two perspectives, empirical and procedural. Empirically, it was thought that what is beautiful is scientifically correct; a correct model does not have to be accurate, and an artificial one might be the correct model. Procedurally, the following were considered important.

- Generalizations beyond human subjectivity are made,
- Evidence (rationale) to demonstrate these generalizations is always prepared,
- Reproducibility is guaranteed, and
- There is refutability against generalizations that lead to saying things are true that are not true.

It is clear that these scientific attitudes underline modern science. The effort to produce better models has had an immeasurable impact on the development of science.

Now that you have understood what a model is and the benefits of thinking about models, let us return to the discussion of computational thinking. Modeling is included in the act of behaving like a computer scientist creating a metaphor for the real world using the mechanisms and operations of a computer. Moreover, the models that generate new information or make predictions are useful to humans.

The major premise of physics is that there is only one correct standard model in this world (or at least that there should be fewer models than there are scientists), but computer scientists consider that scientists, when observing the real world, create models from different perspectives, so there can be as many *correct* models as there are scientists. Undoubtedly, computer science is a rare discipline in which multiple, possibly conflicting models may be allowed. It is the versatility of computers that allows for flexible proper modeling. Therefore, when people model various things and events in the real world from their own perspectives, it is computer science that can provide useful insights, and this is computational thinking.

## 2.4 Virtualization

### *2.4.1 Meaning of Virtualization*

Finally, let us discuss another important concept, virtualization. In Section 1.2.3, we mentioned that the second condition of good computation is to find a more efficient or effective solution to the problem represented by the model after bringing the things and events of the real world into the world of the computational model. This solution is called a procedure or algorithm, and the explicit representation of the relationship between data is called a data structure. Rather than processing related elements separately, it is desirable to group them into a vector or matrix and process them all at once to achieve a readable program or algorithm for easy improvement.

Thanks to long-term research on algorithms and data structures by computer scientists, people can easily access a generic catalog for excellent solutions. If you can model a problem of the real world in such a way that the cataloged algorithms and data structures can be applied smoothly, you can easily obtain lean processing. This is a good computation, and good modeling is the key to good computation. Then, the key to good modeling is virtualization, which is the topic of this section.

### *2.4.2 Does Virtual in Virtual Reality Really Mean Virtual?*

First, we have to clarify the meaning of the word "virtual" because it is used in two major senses. Here, let us ask ChatGPT what "virtual" means. The answer is as follows:

> "Virtual" generally refers to something that exists or occurs in a digital or simulated environment, rather than in the physical world. It can also denote the imitation or representation of something, often created or experienced through electronic or computer technology.

This answer is ambiguous as to whether the original object being simulated or imitated is real or not. In other words, virtual has two major meanings. One is to extract or focus only on the essential or substantial parts of something that already exists in the world, and the other is to make something appear as if it actually exists, including things that do not exist in the world. Thus, the latter is more broadly defined and includes the nuances of fictitious and hypothetical. The former does not include such nuances and only looks at things based on reality.

The term "virtual reality" is sometimes used in the mass media to mean making you perceive and feel as if the fictitious space created by a computer is real, bringing experiences that are impossible in reality. In contrast, the "virtual reality" commonly used in the academic community is defined, for instance, as follows:

> A technology or a general term for a system that creates the essential functions (the essence of reality) of events and phenomena in the real world, even if their appearance and form are quite different from those in the real world, by presenting information generated by a computer to various human senses.

Incidentally, the Japanese Society for Artificial Intelligence suggests that "artificial reality" might be a less misleading term than virtual reality.

Hereafter, the word "virtual" in this book will be used in the sense of virtualization as the reproduction of a situation that is essentially the same in function as things and events in the real world. However, simply reproducing a situation does not benefit anyone, so it is reproduced in a way that is easy for users to handle. In focusing only on logical existence and objects, by removing the physical structure and constraints of the real world, virtualization means the same as abstraction. However, the key point of virtualization is to "reproduce in a way that is easy for users to handle." In reproduction, it absorbs the minor physical differences that realize its function and state and provides only the substantial function. This leads to "ease of use for people."

**Column 9 the Virtual President**

The English word "virtual" can mean "in effect, if not in name." It is said that Bill Clinton, the 42nd President of the United States, once said *"Hillary is the virtual President."* This means that Hillary is the president in effect, and it may be implied that Bill is the nominal President. Hillary is the wife of former President Clinton, and she is said to be a more capable politician than her husband. As you all know, she later ran for president against Mr. Trump and was narrowly defeated.

### *2.4.3 Virtual Memory*

The quickest way to understand virtualization is to look at a concrete example. There is a technique called virtual memory that virtualizes memory (Fig. 2.1). Physical memory is realized by semiconductors and has a fixed capacity such as 512 KB or 1 GB, but the logical address space expressed in the program is 32 or 64 bits wide, the memory space of which is 16 GB or 16 EB, respectively. It is difficult, both in terms of cost and speed, to allocate real memory for the entire logical address space. Accordingly, a technique has been developed that realizes an ideal memory, in which physical memory exists throughout the entire logical address space when viewed from a program, despite using only a small amount of physical memory that covers only a fraction of the logical address space in reality. This is a virtual memory. Specifically, during the execution of a program, the requested logical addresses are sequentially associated with the physical addresses where the actual data exists. Logical addresses that are no longer used are automatically disassociated with their physical addresses. The substantial function that was focused on here is the memory address. From the program's perspective, it can freely be used anywhere in the logical address space, and it can dispose of it without being aware of the reuse of the memory area. This

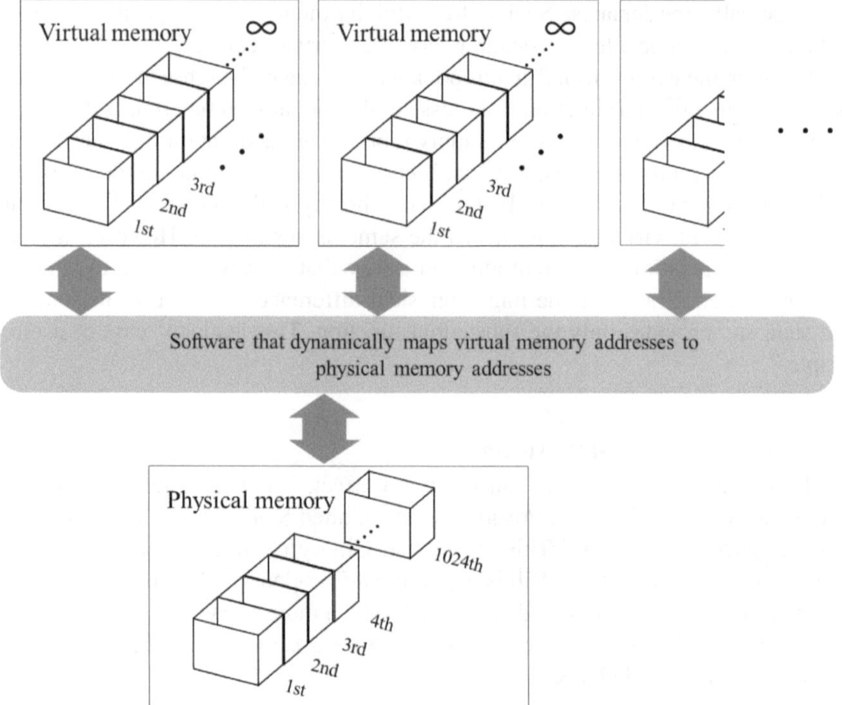


**Fig. 2.1** Mechanism of virtual memory

alone makes it "easy for the user (in this case, the programmer) to handle", yet what is even more amazing about virtual memory is that multiple programs (processes) can use the same logical address at the same time without any problems. In other words, you can write a program without worrying about which part of the logical address other processes are using. Virtualizing the physical memory address and presenting it to the programmer as a logical address makes memory extremely easy to handle.

### *2.4.4 Virtual Machine*

Another example is the technique of virtualizing the CPU, which is called a virtual machine (Fig. 2.2). There are companies like Intel and AMD that manufacture and sell physical CPUs, and they have a variety of products lined up. For embedded use, Raspberry Pi and Arduino are popular. These different CPUs operate in different machine languages and have different functions, so if you want to implement a Python language processing system (compiler, runtime system, etc.), an OS or applications that work the same way on any CPU, you have to create a program in a different machine language for each different CPU. This is a very labor-intensive task. Therefore, it is preferable to set up only one logical CPU and machine language. That is the virtual machine. The virtual machine is implemented in the machine language of different real CPUs, and the Python language processing system, an OS and applications are implemented in the machine language of the virtual machine. The machine language of the virtual machine is also called the intermediate language. While different CPUs may be used, the components running under the virtual machine's machine language can be commonly used. The substantial functions that are focused on here are the functions and machine language of the CPU. This alone makes it "easy for the user to handle", but what is even more amazing about virtual machines is that multiple independent virtual machines can be created on a single physical computer (CPU). In other words, if you prepare a CPU with powerful computing power and launch multiple virtual machines on it, the computing power, memory capacity, and communication capabilities of that CPU can be shared among multiple virtual machines, resulting in a flexible computer environment with little waste. The provision of these virtual machines to users over a network is what is known as the cloud.

Thus, the direct benefits of virtualization are the ability to flexibly divide, copy, integrate, and transform computational resources of a real computer. Generally, resources mean things that are necessary for activities and that decrease when consumed. Computational resources refer to the resources necessary for executing and realizing computations, specifically, time, memory space, communication lines, etc. By virtualizing computational resources, users can adjust the quality and quantity of computational resources at their disposal and freely process them (adjust them for their own purposes).

### *2.4.5 Various Names for Virtualization*

In the various fields and application areas of computer science, the concepts equivalent to virtualization are called by different names. First, in the field of information systems, a group of services further developed from virtual machines is called the cloud (Fig. 2.3). In addition to virtual machines, services that provide users with the functions of specific software or applications are called SaaS (Software as a Service). Furthermore, services that provide users with not only applications but also their development environment are called PaaS (Platform as a Service). And services that virtualize the entire computational environment, including memory and network, and present it to users as infrastructure, are called IaaS (Infrastructure as a Service). In the field of networks, the extraction and commonalization of only the substantial functions of the equipment or processes to be controlled are called standardization and normalization. Standardization and normalization improve interoperability. In the field of software engineering, appropriately dividing a program into independent parts, focusing only on its functions and interfaces, and abstracting (hiding) its implementation is called modularization, componentization, object orientation, etc.,

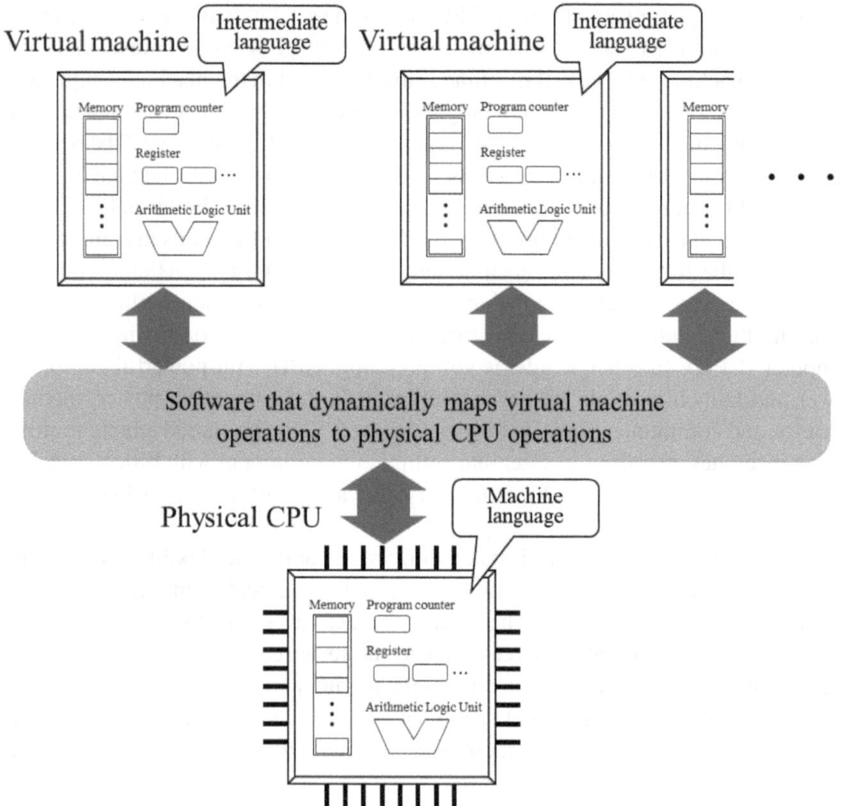


**Fig. 2.2** Mechanism of virtual machine

which correspond to the virtualization of programs. The virtualization of programs both enhances the reusability of the programs and contributes significantly to software design theory, by which large-scale software is accurately and efficiently created by a small number of people.

### 2.4.6 *Other Virtualizations*

Not only in the world of computer science, but also in the real world, you will be able to find many examples of virtualization. In the past, physical entities and the functions, services, and contents provided there were inseparably linked, but by virtualizing them, we produce the trend of separating functions, services, and contents from physical entities and providing users with more essential things, a trend which is accelerating nowadays. Virtualized functions, services, and contents are freely divided, copied, integrated, and transformed according to the purpose and

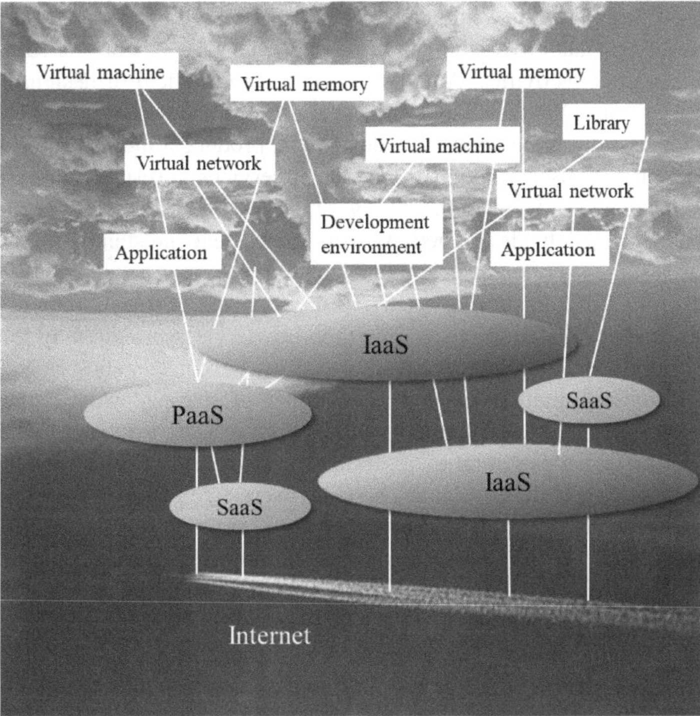


**Fig. 2.3** Cloud = Virtualization of entire computing environment

environment on the provider and/or recipient sides, and are used more appropriately and conveniently. For example, when listening to music, in the past, the physical entity (a vinyl record) and the content (music) were inseparably bound together, and even the playback method was determined and limited. The advent of CDs enabled music to be handled as digital data, freeing it from the constraints of sound quality and the limited playback methods of vinyl records. After that, the sound of music has become digital data, independent of the physical form of a CD, allowing users to download it via the Internet without having to acquire a physical medium. Now, with streaming, only the digital data necessary for playback is transferred on demand.

Human activities are also being virtualized. In society, humans can be seen as integrated entities possessing knowledge, experience, and skills, and digital technology has been separating these attributes from the physical existence of humans. For example, when reading an email, you can understand the sender's message, but you usually do not know where the sender was when they wrote the email. There is also the gig economy, a form of labor market that separates and extracts specific abilities of a human, integrates them, and serves them in a cloud-like manner, which can be said to be a virtualization of human activity.

The wave of virtualization is also affecting the way corporate organizations operate. Companies have emerged that provide an outsourcing service, taking on only the common general affairs and personnel tasks of other companies (for example, ADP (Automatic Data Processing) and Insperity). This can be considered a virtualization of corporate activities. Another example is in the field of design, in which there is a focus on UX (User Experience). Companies and designers have emerged who specialize in thinking about how design can produce a more comfortable and richer experience for users through the use of web services and establishing a comprehensive and extended design methodology. This can be considered a type of virtualization that separates the design activity from the context and standardizes it, while traditional design was often inseparable from context.

Furthermore, virtualization is occurring in the world of transportation, such as taxis, city buses, and shuttle buses. The mobility services provided by traditional taxis, city buses, and shuttle buses were bound by parameters such as route limitations, timetable limitations, and the possibility of sharing, which were tightly restricted by law. In addition, these mobility services were inseparable from their hardware (vehicles, stops, operation/reservation systems, etc.), and as a result, offered very little freedom. If these mobility services were freed from the constraints of the law and separated from the hardware by virtualization, it would be possible to consider a generic mobility service where each parameter can be freely adjusted (Fig. 2.4). As mentioned earlier, the benefits of virtualization include the ability to flexibly divide, copy, integrate, and transform resources. If these general mobility services can be divided, copied, integrated, and transformed, they could provide optimal mobility services according to passenger demand. This corresponds to the concept of MaaS (Mobility as a Service), which has become increasingly common in recent years. It can also be rephrased as the cloudification of mobility services.

The final example of virtualization is the 3D printer, which has revolutionized manufacturing. Traditional manufacturing required tools, materials, blueprints, and a maker with skills, knowledge, and experience, all of which needed to be gathered in one place at the same time. However, the advent of 3D printers has dramatically relaxed these constraints. As a result, anyone can now make certain things that were previously only made by craftsmen or experts. Manufacturing skills used to be scarce, but it will no longer be so. Traditionally, goods had to be mass-produced and consumed in large quantities due to cost and time constraints. However, it is now possible to produce a small variety of goods and consume locally. Dr. Hiroshi Ishii, the associate director of the MIT (Massachusetts Institute of Technology) Media Lab, has proposed the concept of "tangible bits and radical atoms", and the 3D printer could be seen as a universal machine that realizes this concept.

Please keep in mind the concepts of abstraction, modeling, and virtualization that we have explained so far, and try to create metaphors for the things and events in the real world using the mechanisms and operations of computers, as computer scientists do. No doubt you will find that there are many opportunities to enjoy practicing computational thinking in everyday life.

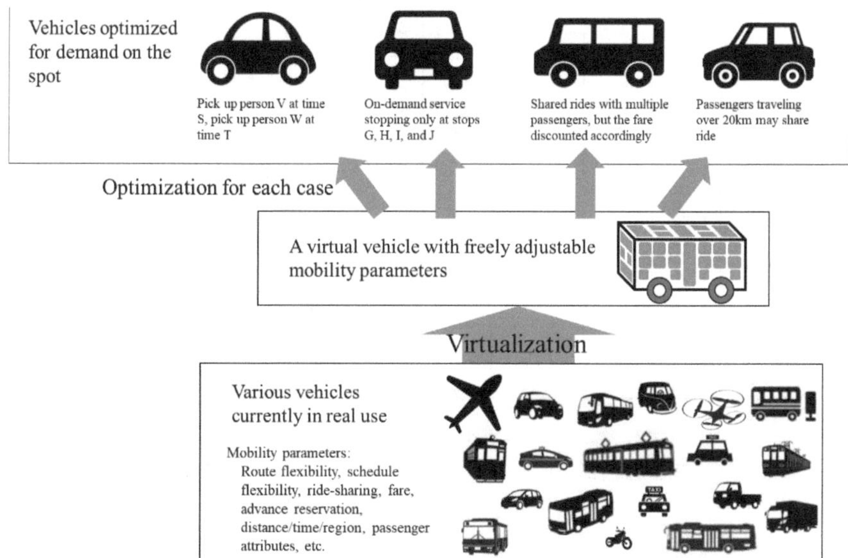


**Fig. 2.4** MaaS as virtualization of entire transportation services

## 2.5 Design

Computational thinking is a way of thinking about how to create new information. The "design" we often hear about is also an activity and field that creates new things, mechanisms, and experiences. So, what is the relationship between design and computational thinking?

Figure 2.5 shows the relationships between various academic and practical fields. The important point here is that computer science and design are orthogonal to other fields. Computer science is now the foundation of all fields. Bioinformatics is a typical example, which is a field that studies biology from the perspective of informatics. Since organisms are made from genes, and these genes hold the blueprint (information) of the body, it is quite natural that these two fields should be combined.

In astronomy, recently, non-optical telescopes such as radio telescopes are used to observe distant celestial bodies. The "photo" of the black hole that made the news in 2019 was taken by the Event Horizon Telescope, which combines the data from eight radio telescopes on the earth and processes it with computers. This is representative of how the data analysis of certain experiments, though the experiments may be physical, cannot be conceived without computers.

Computer software can basically program anything you can think of. However, when applied to fields such as transportation and healthcare, it is necessary to carefully consider human preferences and requirements. That's where design comes in, introducing values of what is desirable and creating mechanisms that are appropriate

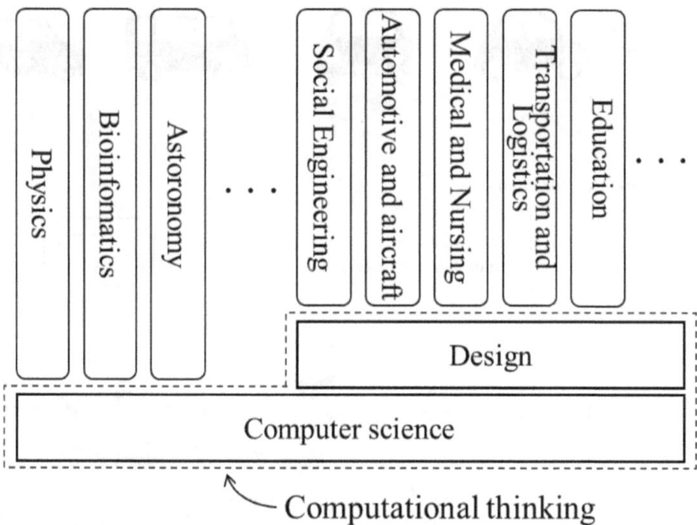


**Fig. 2.5** Relationship between computational thinking and design

for it. Design is the foundation of all engineering. We believe that computational thinking lies in the combination of insights from both computer science and design.

# Chapter 3
# What is Meta?

**Yasuhiro Katagiri**

**Abstract** This chapter explains the concept of "meta," one of the essential concept in mastering computational thinking. As a reminder, computational thinking is defined as "thinking like a computer scientist." Computer scientists use computers as tools to generate new information, and they are constantly considering how to recreate the world's objects and events within a computer. To do this, they must reflect and observe everything from a higher perspective to represent the world's objects and events as a program. This act and concept are called meta. Since meta can be a difficult concept, this chapter will explain it using examples from everyday life. Humans use meta-thinking effortlessly and unconsciously but achieving it with a computer is extremely challenging. This is because computers must also handle self-reference—the act of referencing themselves—since all things and events include data about data, knowledge about knowledge, learning about learning, and thinking about thinking. This chapter will carefully explain both meta and self-reference using examples from language, mathematics, and computation.

John Cage, an American contemporary music composer has a piece titled 4′33″. A performer comes on stage, prepares to start playing, and then remains silent for 4 minutes and 33 seconds without producing any sound. During this time, the audience, who are anticipating and ready to hear music, experience silence instead of actual music.

Through this absence of performance and the resulting silence, the audience is forced to confront the question: what is the very act of listening to music itself. It seems that 4′33″ is designed to produce this kind of effect. Some people refer to this experience as meta-listening, meaning listening to the act of listening itself.

One of human's distinguishing characteristics, when compared to other living beings, is our exceptional ability to learn. When we learn something—learning to speak a foreign language, learning to program, or learning how to cook—there is

Y. Katagiri (✉)
National Institute of Advanced Industrial Science and Technology, 2-4-7 Aomi, Koto-ku, Tokyo, Japan
e-mail: y.katagiri@aist.go.jp



H. Nakashima and K. Hirata (eds.), *Computational Thinking*,
https://doi.org/10.1007/978-981-95-5962-6_3

a specific subject or "something" to be learned. In contrast, the concept of meta-learning has recently been gaining attention. Meta-learning involves reflecting on what and how we have learned during or after acquiring knowledge of "something" (like programming). By doing this reflection, one can improve and enhance their methods of learning, leading to better learning outcomes. Meta-learning is considered an advanced way of learning. It goes beyond learning specific subjects and helps one acquire highly generalizable learning strategies that can be applied across various domains. Moreover, it fosters an autonomous approach to learning.

When we engage in activities like listening to music or learning to cook, these are actions targeting specific objects, referred to as object-level actions. On the other hand, actions like listening to the act of listening itself or learning about the act of learning are actions abstracted to a higher degree. They are referred to as meta-level actions. Conceptually, it is even possible to imagine meta-meta-levels, such as listening to the act of listening to listening or learning about the act of learning about learning.

## 3.1 Meta in Language

Compare the following two sentences.

(1) Snow is white.
(2) "Snow" is a noun.

Sentence (1) describes a fact about a type of meteorological object "snow" it has white color, a fact about physical world. Sentence (2), on the other hand, describes a fact about an English word "snow" that it is a noun, a fact about English grammar. The same word "snow" has a completely different meaning in sentences (1) and (2). In the former case, a word talks about a physical entity, so it belongs to object-level language. In the latter case, a word talks about the word itself. So it belongs to a meta-level language. This contrast is called use and mention dichotomy in linguistics,

The distinction between object-level and meta-level also exists not only in human language but in the language of computers, specifically programming languages. Ordinarily, programs are created to enable computers to perform specific tasks. Whether it is providing services via a smartphone app, controlling characters in a game, or driving a car autonomously, all these functions rely on programs. These programs are designed to address problems that exist outside the computer, and the language used in such programs is referred to as the object-level language. In contrast, like sentence (2) in English, there exists a type of languages used to describe what a programming language itself is, the grammar that distinguishes correct programs from incorrect ones, or the specification of a programming language in general. They are the meta-level programming languages. Meta-level programming languages are used to define the rules and structure of programming languages themselves. These programming language specifications are utilized in building systems like compilers or interpreters to ensure they function as intended and without errors.

You might think it should be possible to make the differentiation between sentence (1) and (2) with the use or non-use of quotation marks. But, removing quotation from (2) retains the same meaning as in (3).

(3) Snow is a noun.

Human languages are tolerant enough to allow object-level language and meta-level language to be one and the same.

## 3.2 Self-reference

Philosophers sometimes (half-jokingly) claim that philosophy as an academic discipline is superior to all other academic disciplines. Their logic goes something like this: all academic disciplines have their own object of studies, mathematics studies mathematical objects, physics studies physical world, economics studies economic phenomena. But they don't take up themselves as their study subjects. Mathematics doesn't ask a question of what mathematics is. That inquiry would belong to the philosophy of mathematics. The situation is the same regardless of whether the subject is mathematics, physics, or economics. Only philosophy would consider the question of what philosophy is. This shows that philosophy is superior to all other academic disciplines. Superiority argument aside, if the description that academic inquiry in philosophy can be directed to philosophy itself is true, then philosophy as an academic discipline can be said to have a self-referential circular structure.

Look at the sentence below.

(4) This sentence consists of six words.

If you count the words, sentence (4) does contain six words. So, what sentence (4) claims is correct, or we might say sentence (4) expresses a true proposition. Different from the most cases in standard language uses, the claim sentence (4) makes is not about some language external objects. Sentence (4) talks about a property of sentence (4) as a language expression, namely it consists of six words. Therefore, sentence (4) is a meta-level language. Sentence (4) has a self-referential circular structure. The expression "this sentence" in sentence (4) makes a self-reference to the sentence (4) itself. The characteristics of human languages are that there is no strict distinction between object-level language and meta-level language, as was shown in sentence (3), making it possible to have this structure.

You can find self-reference in legal domains, e.g., the Japanese Constitution. There is a clause in Article 59 of the Japanese Constitution that states: "A bill becomes a law on passage by both Houses, except as otherwise provided by the Constitution." If we think ordinary laws are to stipulate the rules that regulate social activities, the Article above of the Constitution stipulates a meta-rule to regulate ordinary laws. On top of this, Article 96 of the Japanese Constitution states: "Amendments to this Constitution shall be initiated by the Diet, through a concurring vote of two-thirds or more of all the members of each House and shall thereupon be submitted to the people for ratification, which shall require the affimative vote of a majority of all votes cast thereon, at a special referendum or at such election as the Diet shall specify. Amendments when so ratified shall immediately be promulgated by the emperor in the name of the people, as an integral part of this Constitution." As the use of expression "Amendments to this Constitution" shows, the Constitution of Japan is completely different from ordinary laws in that it has a self-referential circular structure by containing rules that determine itself. If we regard the Constitution as a meta-rule that defines ordinary rules, Article 96 is a meta-meta-rule that defines the Constitution meta-rule.

In the examples above, expressions like "this sentence" or "this constitution" are used to perform self-reference. While a noun like "sentence" always refers to the general category of sentences regardless of where it is used, pointing to a specific sentence in the text and saying "this sentence" refers to that particular sentence. If you point to a different sentence and say "this sentence," it then refers to a different one. Expressions such as "this sentence," "this," "that," "it," "I," "you," "now," and "here" share a common property: they indicate specific objects contextually determined by the circumstances in which they are used. These are known as indexical expressions. In the case of examples like sentence (4) or the constitution, indexical expressions are used to refer directly to themselves, achieving explicit self-reference.

Self-reference can also be realized in more subtle, indirect ways. In the history of science, many laws and concepts have been named after individuals, such as Newton's laws of motion or Ohm's law. Those concepts associated with names of individuals are called eponyms. The purpose of eponyms is likely to honor the original proponents of important concepts by naming these concepts after them, ensuring that their contributions are remembered.

However, upon closer examination, it is often found that the person whose name is associated with an eponym was not actually the first to propose the concept. This seemingly surprising finding is known as Stigler's Law of Eponymy. What is interesting is that Stigler's Law of Eponymy itself is an eponym named after "Stigler," yet the law was first proposed by someone other than Mr. Stigler. Stigler's Law of Eponymy applies to itself, making it a valid example of the law through indirect self-reference.

In philosophy, there is a position known as relativism. Simplified, it argues that there is no absolute truth in the world; truth is always relative to the perspectives and values of the people addressing it. A well-known counterargument to this claim points out that, if relativism is true, then the claim of relativism itself cannot be absolute and is relative to the perspectives and values of specific individuals. Thus, it cannot be universally correct. Indirect self-reference provides a powerful means for making a counterargument. Another example would be a story about a country's Health minister who, in encouraging citizens to live with humor for better health, delivered the suggestion with a stern, humorless expression, evoking laughter instead. This episode also highlights the power of indirect self-reference.

## 3.3 Meta in Computation

At the beginning of the twentieth century, mathematician David Hilbert proposed 23 problems as significant challenges in mathematics. Hilbert's problems are said to have shaped the subsequent development of mathematics. Among these, the second problem involved formalizing arithmetic, which deals with addition and multiplication of natural numbers. Hilbert believed there was confusion surrounding the foundations of mathematics at the time. To eliminate this confusion, he proposed an idea to reinterpret mathematics as a formal system. In this system, only propositions derived mechanically through finite applications of inference rules from axioms would be accepted as proven. As a relatively simple example, he chose the formalization of arithmetic. However, contrary to Hilbert's expectations, a mathematician Kurt Gödel published a paper in 1931 in which he proved the impossibility of such formalization, thus arriving at a negative conclusion. This result is known as Gödel's Incompleteness Theorems, an eponym named after Gödel.

When constructing a formal system, it is essential to ensure that no contradiction arises, meaning that both a proposition $p$ and its negation $\neg p$ cannot both be provable. Such a system is called consistent. On the other hand, a weak system where neither $p$ nor $\neg p$ can be derived is insufficiently powerful. A system is considered complete if, for every proposition $p$, either $p$ or its negation $\neg p$ is provable. Gödel demonstrated two results: In any consistent formal system of arithmetic, there will always exist a proposition $p$ such that neither $p$ nor $\neg p$ is provable (First Incompleteness Theorem). The consistency of the formal system itself cannot be proven within the system (Second Incompleteness Theorem). In other words, it is impossible to create a consistent and complete formal system even for arithmetic, which was thought, by many mathematicians including Hilbert, to be fundamental and straightforward to handle rigorously.

Gödel used the idea of indirect self-reference at the meta-level to prove his incompleteness theorems. First, he assigned unique numbers to all propositions. Propositions about numbers, such as "2 is even," can be expressed as sentences formed by a finite sequence of symbols. By first assigning numbers to each individual symbol, such as "2," "is," and "even," and then devising a method to combine them into a single number, it is possible to uniquely assign a number to each entire proposition. These assigned numbers are called Gödel numbers, and the Gödel number of a proposition $p$ is denoted by $\ulcorner p \urcorner$. With the idea of Gödel numbers, object-level propositions about numbers, like "2 is even," and meta-level propositions about those propositions can be treated equally within the formal system of arithmetic. This has the same structure as in language, where by using parentheses, one can express object-level statements like "Snow is white" and meta-level statements like "'Snow' is a noun" without distinguishing them. At the meta-level of propositions, one can, for example, consider properties such as whether a particular proposition is provable within the formal system of arithmetic, specifically whether a given number is the Gödel number of a provable proposition.

Next, by employing a technique called diagonalization, indirect self- reference at the meta-level can be realized. Specifically, for any property $A(n)$ of a number $n$, it is possible to construct a proposition $D$ such that within the formal system, the equivalence of $D$ and the proposition $A(\ulcorner D \urcorner)$, which states that the Gödel number of $D$ possesses the property $A$, can be proven. By substituting "not provable" (the negation of "provable") for this property $A$, it is established that there must exist a proposition $G$ within the formal system such that $G$ is equivalent to the statement "the proposition $G$ is not provable." In essence, $G$ makes a claim about itself:

(5) "This proposition is not provable"

If $G$ were provable, then it would simultaneously claim its own unprovability, creating a paradox. Consequently, neither $G$ nor its negation can be proven, demonstrating that formal arithmetic systems are inherently incomplete.

Though the logical structure may seem complex, the core idea lies in using Gödel numbering and diagonalization to achieve indirect self-reference at the meta-level. As a side note, the diagonalization technique used here is fundamentally the same as the one employed by a mathematician Georg Cantor when he showed that the set of real numbers is larger than the set of natural numbers.

This development from Hilbert to Gödel was underpinned by the concept of mechanically applying logical operations. To further clarify this concept, a mathematician as well as a logician, Alan Turing, who is counted as one of the founders of computer science, proposed an abstract computational machine, later called the Turing machine. This too is an eponym. Turing is also known for his works in cracking the German codes, an achievement that played a significant role in the eventual victory of the Allies in World War II.

A Turing machine consists of a tape with infinitely many cells arranged in one dimension and a head that reads, writes, and moves left or right on the tape. The machine transitions between multiple states, and each cell on the tape can hold a symbol from a predefined set. The machine's behavior is precisely defined by a set of rules that dictate mapping from the two items, the current machine state and the symbol read from the tape, to the three items: the next machine state, the symbol to write on the tape, and the movement of the head. When a Turing machine can be appropriately configured in such a way that starting from symbols on the tape representing two numbers $n$ and $m$ and operated according to the rules, it results in symbols on the tape representing $n + m$, then we can say that a Turing machine for addition is constructed. Thus, addition is computable by a Turing machine. Turing machines are conceptual devices, and they need not be constrained by practical considerations. It is perfectly fine to choose a not-so-practical way of representing numbers, e.g., placing $n$ consecutive "1"s on adjacent cells to represent the number $n$.

Once computability was formally defined by Turing machines, it became possible to frame questions such as computability of the irrational number $\pi$ as a question whether the digits of the irrational number $\pi$ can be sequentially written on the tape by some Turing machine. Similarly, for Hilbert's mathematical problem on formal systems, Entscheidungsproblem (decision problem) in mathematics—determining whether a proposition is provable in a given formal system—was shown to be undecidable in general.

The most important idea concerning Turing machines is the concept of Universal Turing Machine. If it is possible to encode the operational rules of a Turing machine as symbols on a tape, it becomes possible to create a universal Turing machine that can simulate the behavior of any other Turing machine by reading its encoded description. Let $p_T$ represent the encoded description of a Turing machine $T$. When the original Turing machine $T$ is given an input $x$ on its tape, let the universal Turing machine $T_U$, starting from $p_T$ and $x$ as inputs on its tape, simulate the original Turing machine $T$ by recreating $T$'s operation steps by referring to both $T$'s encoded description $p_T$ and original input $x$. Turing demonstrated that such a universal Turing machine is constructible. In essence, $p_T$ corresponds to a program. This idea of a universal Turing machine was realized in the form of electronic circuits by John von Neumann, who is a mathematician, a physicist, and a computer scientist, and others as stored-program computers. Those machines eventually lead to the creation of modern computers.

The description $p_T$ of a Turing machine $T$ plays a similar role to Gödel numbers. Following a similar process in Gödel numbering in formalization of arithmetic, Turing machine description $p_T$ can also be assigned a number. Let's call this number the Gödel number of the Turing machine. The construction of the Universal Turing Machine provided Turing machines and hence stored-program computers with the powerful capability to treat both the object level: problem domains expressed by the input $x$, and the meta-level: computational processes expressed by the description $p_T$ as equally manipulable entities.

The ability to mix object-level and meta-level concepts enables indirect self-reference at the meta-level. Let us consider what is called the Turing Machine Halting Problem. This asks, given a Turing machine $T$ and input $x$, whether we can determine $T$ will halt on $x$? This problem effectively asks whether a Turing machine can be constructed that decides, when a Turing machine is given, whether it finishes its computation and stops with a result or it does not stop, and computation keeps going indefinitely. In parallel to the case of Gödel Incompleteness Theorems, a negative answer can be obtained for this problem through diagonalization and meta-level indirect self-reference.

Let's assume that it is possible to construct a halting machine $T_H$ that can solve the Turing Machine Halting Problem. $T_H$ takes as input the Gödel number $\ulcorner T \urcorner$ of a Turing machine $T$ and an input $x$ to $T$, and returns "yes" if $T$ halts on $x$ and "no" otherwise. We can derive a contradiction using diagonalization. Construct a new machine $T_D$ based on $T_H$ that, when given the Gödel number of a Turing machine $T$, $\ulcorner T \urcorner$, it will answer "yes" if $T$ does not halt on $\ulcorner T \urcorner$ (as input $x$ to $T$) and "no" if it halts. Then, it is easy to see that feeding $\ulcorner T_D \urcorner$ as input to $T_D$ itself leads to a logical contradiction. When $T_D$ is given its own Gödel number $\ulcorner T_D \urcorner$, it stops and answers yes, if the computation of $T_D$ does not halt on $\ulcorner T_D \urcorner$, and "no" if it does halt, hence a logical contradiction. Thus, $T_H$ does not exist, and the halting problem is undecidable. This argument hinges on indirect self-reference at the meta-level.

We can find indirect self-reference at the meta-level in modern programming languages. Some programming languages, like Lisp language, provide users with a mechanism of direct interpretive execution of source codes. Users are spared from compiling steps, e.g., transforming source codes into executable codes, before executing their programs. An interpreter for Lisp language, which controls the interpretive execution of Lisp source code, can be written in Lisp language itself. This type of interpreter is called a meta-circular interpreter. If you write a Lisp interpreter in Lisp, in order to run a Lisp program, you will need another interpreter for the interpretive execution of Lisp interpreter itself, and that interpreter needs another layer of interpreter, ad infinitum. All of these interpreters can be the same, hence circular, but another mechanism is necessary to actually run the code. So, it is not very practical to have a meta-circular interpreter as such, but it is useful to conceptualize computation and to provide a foundation in designing programming language systems.

## 3.4 Meta of Thinking

The seventeenth century French philosopher René Descartes famously stated, "I think, therefore I am." If thinking about tomorrow's weather or solving a math problem constitutes object-level thought, then reflecting on the act of thinking itself, as Descartes did, represents meta-level thought. Introspection, or self-examination, is closely tied to meta-level thinking.

Who, then, is having the thought "I think, therefore I am"? The typical answer would be "I." But this seemingly innocuous answer that thought belongs to "I" contains two intriguing problems: infinite regress and subjectivity. If "I" think that "I think therefore I am." then "I" must also think that "I think that I think therefore I am." Still, there must be "I" who think "I think that I think that I think therefore I am," and so on indefinitely. This structure mirrors the infinite regress argument against Artificial Intelligence. Realizing intelligence in a computer requires an intelligent module, a homunculus, which can handle information in the computer intelligently, and realizing this homunculus requires another subordinate intelligent module, a homunculus, ad infinitum. Because of this infinite regress, Artificial Intelligence is impossible. The claim that meta-circular Lisp interpreters cause infinite regress and cannot execute Lisp code by themselves has a similar flavor. The presumption that thought belongs to "I" seems to conceal infinite regress.

Descartes considered the thought "I think, therefore I am" to be private, which belongs only to "I" and is inaccessible to others. For "I" it is directly accessible and constitutes the most certain of all knowledge. Pursuing further the idea of private and direct subjectivity leads to the concept of qualia. The blue color of the sky I am now looking at is a part of MY subjective experience, and even if another person is looking up at the same sky at the same time, she could have a completely different experience, which might possibly be equivalent to the perceptual experience I would have when I were looking at something red. I know dogs have superior olfactory senses than humans. Despite the knowledge I cannot know how my dog perceives the smell in our room. Humans (and some of animals) have such subjective experiences, qualia. With advancements in neuroscience, might it become possible in the future to assess the color I am looking at more accurately than myself? I: "I think I am looking at the color blue." Brain scientist: "No, you are wrong."

We have explored mechanisms of indirect self-reference at the meta-level in both language and computation, such as quotation or Gödel numbers. We had a glimpse of the existence of lots of possible ways to cause indirect self-reference at the meta-level from such examples as Stigler's Law of Eponymy or Health minister's humor. Both language and computation are constructed on top of shared systems of clearly defined symbols. Subjective experiences, on the other hand, are not always accompanied by clear expressions in language. Meta-level reflexivity permeates areas like music, where pauses and silence are as meaningful as notes. I mentioned John Cage's work 4′33″ at the beginning. Not as extreme, inserting a silence during a musical play by interrupting performance is a well-known technique employed by a lot of musicians, including an eighteenth century composer Joseph Haydn. A Jazz musician Miles Davis once said, "Don't play what's there, play what's not there." Japanese music emphasizes the importance of a pause ("Ma"). In philosophy, concepts like qualia and infinite regress challenge our understanding of thought itself. When it comes to Meta of Thinking, I believe that there are still a lot of rooms to explore possible mechanisms of indirect self-reference at the meta-level. Unraveling the diverse mechanisms underlying meta-level thinking remains an open frontier.

The subject of this chapter may be somewhat of an outlier in this book, which aims to provide easily accessible expositions of concepts in computer science such as programming with reference to everyday objects or events. It is beyond my hope if readers are interested in the broader scope of computational thinking including topics covered in this chapter. I conclude with this meta-level comment.

# Chapter 4
# Life and Computational Thinking

**Noyuri Mima**

**Abstract** This chapter is an essay in which the author, a computer scientist and education scholar, credits her ability to readily embrace computational thinking upon first encountering it to formative personal experiences from her childhood. She reflects that her participation in the math club during junior high school, where she delighted in mathematical puzzles and logical reasoning, together with the later experience of exploring LOGO programming at university, gradually nurtured an approach very similar to what is now called computational thinking. Because the author has consistently explained abstract ideas by drawing analogies to everyday objects and concrete activities such as cooking, she once again frames computational thinking through this familiar lens. The entire process of cooking—setting a goal and breaking it down, designing and refining a recipe, planning and scheduling, preparing ingredients, executing each cooking step in order, and finally cleaning up—is interpreted in detail from the perspective of computational thinking, making each step relatable and vivid. She encourages readers to experiment with introducing computational thinking perspectives into the ordinary activities they already know well. Finally, at the end of this chapter, from a broader educational standpoint, the author reflects on the connection between what children of the future should learn and the cultivation of computational thinking, and she offers a carefully considered proposal for how such thinking can best be learned.

## 4.1 Thinking from Personal Experience

Professor Eiichi Osawa, a colleague of mine at Future University (FUN), introduced computational thinking in a column on the website of the Information Processing Society of Japan (IPSJ) in January 2014 as "Encouragement of 'Computational Thinking,'" and then Hideyuki Nakashima, one of the editors of the book and former president of Future University, translated Jeannette Wing's essay and introduced it in the IPSJ Journal in June 2015.

---

N. Mima (✉)
Future University Hakodate, Hakodate, Hokkaido, Japan
e-mail: noyuri@fun.ac.jp

 
H. Nakashima and K. Hirata (eds.), *Computational Thinking*,
https://doi.org/10.1007/978-981-95-5962-6_4

Wing's ideas has had an impact on a wide range of educational circles. For example, in the US and Europe, computational thinking has been discussed in K-12 education, and many books, textbooks, and curricula have been developed and implemented. On the other hand, the term "programming-like thinking" has become widespread in Japan, and it has been confused with computational thinking.

To discuss our version of Computational Thinking, a series of colloquium was held six times at Future University Hakodate. The title of my talk at the colloquium was "Thinking about Computational Thinking from Personal Experience" on May 12, 2017. In this chapter, I'd like to start with that talk and introduce some things I've been thinking about since then. Let's do computational thinking, which computer scientists and a few others consciously enjoy, for everyone can benefit from it.

FUN is home to researchers in computer science and a variety of other fields such as cognitive science, mathematical science, artificial intelligence, cognitive psychology, educational technology, design, and communication, and our educational and research activities have an interdisciplinary orientation. There are people who understand and can practice computational thinking, people who do not think they can, but they actually do it, and people who think it's irrelevant. In the "Computational Thinking Colloquium," we will not only introduce the computational thinking discussed in Europe and the US but also reexamine what computational thinking is in the first place. We thought that the process of thinking and creating a book together, to publish it from the FUN Press, would be meaningful in itself.

The computational thinking I think of is represented by two experiences. One is the programming language LOGO and the other is about cooking.

To make this publishing project a success, I thought it was necessary to do it with the idea of benefiting all parties. The colloquium presenters could organize their thoughts. The audience can learn new things, be inspired to think and participate in discussions. And most importantly, people who have read the book will be glad to know about it.

First, here is the definition that Wing gave at the lecture I attended at the University of Tokyo in 2017.

> Computational thinking is the thought process involved in formulating a problem and expressing its solution in a way that can be effectively executed by a computer (human or machine).

As the discussion has expanded, Wing has been gradually changing the definition, and currently emphasizes four things in computational thinking:

> First, computer science is different from mathematics or engineering in that there is software, and software can do many things.
>
> Second, the characteristic ideas of computational thinking are algorithms and data structures, models and simulations, and abstraction.
>
> Third, computational thinking is practical beyond computer science in economics, law, life sciences, archaeology, journalism, humanities, and social sciences.
>
> Fourth, the weakness of computational thinking is that it is difficult for non-computer science majors to understand.

The fourth one is particularly interesting. Wing herself is aware that it is difficult for people in other fields to understand. I wonder if this is because she has seen many examples of computational thinking being misunderstood in the education of computational thinking that is being provided in many places. In the previously mentioned lecture, she said she only advocates and promotes computational thinking but does not participate in education. Instead, she noted that it was something that educators should think about, and she would not interfere.

As I have repeatedly explained in this book, the problem-solving process using computational thinking has the following characteristics:

- Organizing data analytically and logically
- Modeling, abstracting and simulating the data
- Formalizing the problem so that a computer can handle it
- Identify, test, and implement possible solutions
- Automate the solution using algorithmic thinking
- Generalize and apply this process to other problems.

## 4.2 Originated in Junior High School

Computational thinking is something that I am very familiar with. This is not only because I majored in computer science at university, but I look back and see that it had started before that.

I came from an integrated junior high and high school for girls and was a member of the mathematics club. Despite the fact that both junior high and high school students were involved in the club, there were only a few members at that time, and I remember that I became the head of the club in my third year of junior high school. I was recruiting juniors during the school festival because I knew that if no new members joined the club in the future, it might be closed. When I asked my former teacher at the alumni reunion the other day, I was very disappointed to hear that the math club was closed shortly thereafter.

In those days, we used to solve mathematical puzzles, make our own puzzles, write scripts that included trigonometry and infinity, make puppets and put on puppet shows, and read programming texts written in English on the small Olivetti computers in the science and mathematics teachers' rooms, and make little games. At that time, I liked to read about mathematical puzzles and games written by Martin Gardner, a mathematician and author, which was serialized in "*Nikkei Science*" (Japanese edition of Scientific American magazine), and its compiled version "*Bessatsu Aha!* "(extra edition "Aha!"). I also liked the Japanese translation of "*Pillow Problems*" by Charles L. Dodgson, the mathematician and author of Alice in Wonderland; and logical puzzle books such as "*Sophistical Logic* (in Japanese)" by Akihiro Nozaki, the Japanese mathematician. What emerges is a picture of junior high and high school students who enjoy mathematical and logical thinking.

When I was a freshman in high school, my club advisor took us on a tour of IBM Japan's headquarters, which was located near the school, and it was the biggest turning point in my life, leading me from mathematics to computer science. I still have the admission ticket from that time. I still have the admission ticket from that time, with the date "Feb. 08, 1977" engraved on it.

After that, I entered the University of Electro-Communications, the only university in Japan with a computer science department at that time. The club activity was contract bridge, which is a favorite game among computer scientists. From first-year students to graduate students and even assistants and professors, we would gather at lunchtime and after school to play bridge while making small talk.

When I was in my third year of college, I was offered a part-time job as a programmer. The job was to port LOGO, an educational programming language developed at the Massachusetts Institute of Technology (MIT), to run on Japanese computers. At the time, LOGO was running on an Apple IIe 8-bit machine. LOGO was a list-processing language similar to LISP, unlike Pascal and FORTRAN, which I had used before. When I met LOGO, I realized that computers could be useful in education, and I researched and developed new educational methods and content.

Seymour Papert, the inventor of LOGO, demonstrated through observation that LOGO could be a powerful thinking tool for developing new thinking skills in children. This idea is the root of Wing's computational thinking.

One of the features of LOGO is the programming method in which a "turtle" cursor appears on the screen, and the user commands it. It is called "turtle graphics" because the picture is drawn as the trajectory of the turtle's movement. The turtle exists not only in the monitor but also as a hemispherical robot with a diameter of about 20 cm, drawing pictures on the floor (Fig. 4.1). Another feature of the robot is that it can easily handle words such as strings. A simple command can run expressions similar to the "natural language" we use in our daily lives (its antonym is an artificially created language, such as programming languages). These two features help us to understand computational thinking.

## 4.3 Tools for Thinking LOGO

Papert was a mathematician who began collaborating with Jean Piaget, a Swiss developmental psychologist who sought to incorporate mathematical analysis into his theories of development. He then moved to MIT, where he started collaborating with Marvin Minsky, the father of artificial intelligence. He developed LOGO, a programming language for children that combines artificial intelligence, mathematics, and developmental psychology. In his book 3, Papert wrote about his childhood fascination with gears: "I was fascinated by the movement of gears, and as I watched them intently for a long time, I realized that different kinds of motion could be created depending on their size and combination." He visualized multiplication tables through the motion of gears. When he encountered a two-variable linear equation at school, he immediately thought of differential gears (a device that combines gears with different speeds to produce different rotations). The gears became a "model" for his thinking, a tool to bring abstract concepts into his mind and help him understand

**Fig. 4.1** A vintage LOGO turtle robot in the author's lab

them. Although gears happen to be a mathematically excellent tool for thought, he wanted to give a good tool for thinking to children who have not encountered such a tool, which led him to LOGO.

Teaching a turtle how to move, or having a turtle draw a picture, is programming in LOGO. For example, to teach a turtle how to move, you need to know to teach a turtle how to move.

forward *distance*

back *distance*

right *angle*

left *angle*

To deepen our understanding, let's explore some examples together.

**Question 1**: What kind of program should I write to draw a square with a side of 50? The answer is (Fig. 4.2).

forward 50

right 90

forward 50

right 90

forward 50

right 90

forward 50

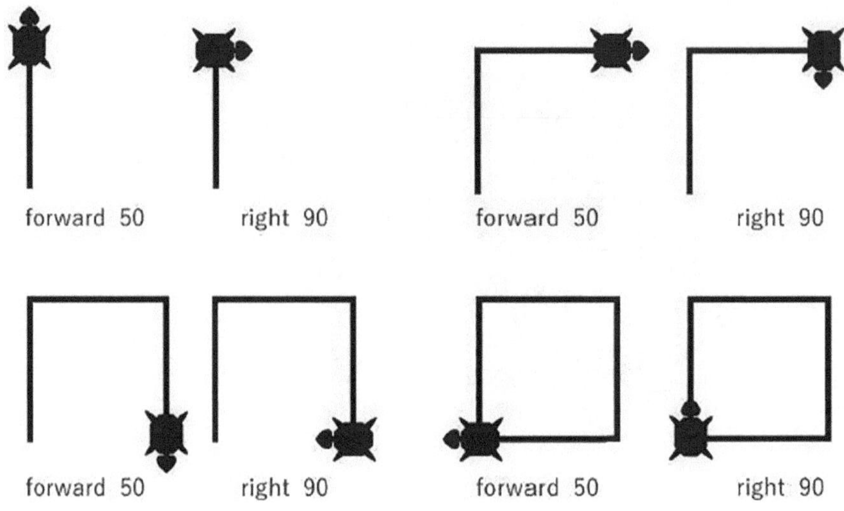


**Fig. 4.2** Let the turtle draw a square

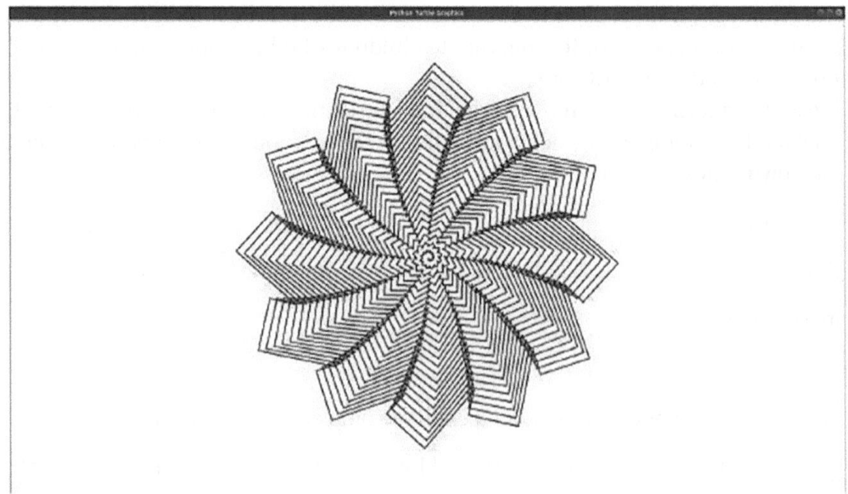

**Fig. 4.3** An example of a diagram you could draw with LOGO0F0F[1]

right 90

This can be written in one line using the "repeat" command.

repeat 4 [forward 50 right 90].

If you program it further, you can write a complex diagram like Fig. 4.3.

[1] https://atlemgw.artstation.com/projects/e0mdxw.

You can also teach the turtle a "word" by defining a function called SQUARE. The definition is as follows:

```
to SQUARE
repeat 4 [forward 50 right 90]
```

**Question 2**: How should I write a program to draw a circle?

It is a little different from the usual drawing of a circle, which is to specify the radius. Let's start with drawing a square and notice that the circle drawn in LOGO is a pentagon, a hexagon, a dodecagon, and so on, with the variable N being the number of sides of the square (4). As N increases, the N-gon becomes closer to a circle.

LOGO allows you to draw complex movements and pictures by combining simple commands. You can draw regular polygons, spirals, and other fractals that are naturally recursive (Sect. 1.5) or iterative. The famous Sierpinski gasket (Fig. 4.4), which is a type of fractal (see Column 10) and consists of numerous self-similar triangles, can also be easily drawn with LOGO.

Furthermore, moving the turtle robot involves a physical sensation, as you try to move like a turtle. The process of synchronizing one's own bodily movements with those of the turtle promotes a deeper acquisition and understanding of geometric concepts. Papert calls this "body-syntonic reasoning". The physicality of LOGO can be considered one of the essential features in teaching and learning.

**Column 10 Fractal**

A problem arises when one tries to determine the precise length of a coastline. Although it may appear straight on a map, enlarging the view reveals intricate curves, and further magnification exposes jagged edges at the granularity of

**Fig. 4.4** The Sierpiński gasket

sand grains. Zooming in further, one can even observe individual molecules. The more closely the view is magnified, the longer the coastline appears. This phenomenon exemplifies a fractal—a concept introduced by mathematician Benoît Mandelbrot. Fractals are characterized by self-similarity: enlarging a portion reveals a shape that mirrors the whole. The figures below illustrate the artistic fractal pattern of the Mandelbrot set.

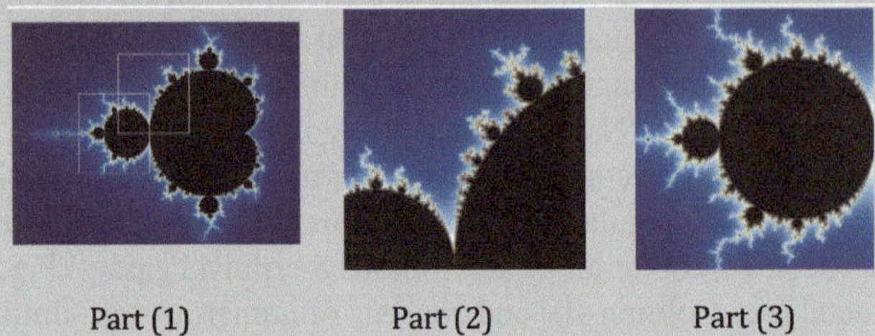

Part (1) Part (2) Part (3)

These works are adaptations of an image by Wolfgang Beyer from https://en.wikipedia.org/wiki/Mandelbrot_set, used under the **CC BY-SA 3.0**

Parts (2) and (3) are magnified views of part (1). In part (2), it becomes evident that the apparent length of the coastline increases with the degree of magnification. Part (3) presents a smoother, self-similar version of part (1).

## 4.4 Thinking with Food

Next, let's think about computational thinking through another experience, cooking. A recipe for a dish can be thought of as a kind of algorithm. I have always used cooking as an example to explain things. This may be because cooking is a very familiar activity to me.

After graduating from university, I worked for an international computer manufacturer, where I was involved in artificial intelligence education and the development of teaching materials. In the textbook, I also used cooking to explain backward reasoning and forward reasoning. Backward reasoning is when you decide what you want to cook or eat and then determine what you need. Forward reasoning is to recall what is in the refrigerator and think about what kind of dish can be made.

The January 8, 1990 issue of Nikkei PC magazine featured a special article entitled "PCs at Home Are Interesting." This was probably because there were few people using computers in everyday life. In that interview, I explained how I structured my recipes using HyperCard. It was not so much programming as system design. On the top page, there is a matrix with pictures of animals. The rows contain the main proteins: pork, beef, chicken, and fish, and the columns contain the words Japanese,

Western, and Chinese. Each cell has a total of 12 buttons, multiplied by 4. Clicking on one of them will bring up, for example, a Chinese menu with fish.

If there is another Chinese menu with fish, an arrow will appear to go to the next menu. Since it is a hyperlink, it is possible to go from a Chinese fish menu to a Western one. The main protein, vegetables, and other ingredients are linked to different menus. This reflects the idea of a home cook who wants to use leftover ingredients efficiently without waste. Of course, you can also search by dish name. There are both backward and forward links, as mentioned above. It doesn't include the cooking instructions, which should be included in every recipe. I recorded my repertoire, listing only the necessary ingredients, seasonings, and tools. It is an auxiliary storage device to ensure I don't forget to shop for ingredients.

You could say it's an application that helps you think about what to make. It's a system for me.

The other things that impressed the reporter in the interview were a household account note and a basal body temperature chart using simple spreadsheet software. I started keeping a household account note when I was a university student, and I still keep it today, which means I have been keeping it for 40 years. The original purpose of a household account book is to review past spending and use it as a reference for the future. But it also is helpful when I want to know how much I have spent in each category.

The basal body temperature chart is probably not very familiar to men. So I recommend that women in their late teens start taking it every morning. Nowadays, there are thermometers and wearable ones that record data directly to a smartphone or PC, making it hassle-free.

At the time of this interview, only analog "gynecological thermometers" were available, with a fine scale showing two digits after the decimal point. I had to rub my eyes sleepily every morning to read the scale, write down the values, and write them on a small graph paper with a small scale, which was specially designed for recording basal body temperature. After a few days, I gave it up and input the data into a spreadsheet. When I went to see the gynecologist at the university hospital, I printed out a graph of the data and brought it with me. My doctor was so surprised that he showed the printed chart to the nurses and doctors around him, saying, "There is an amazing patient who made this." It was not a big deal at all for me, and I just wanted to make things easier.

All of these examples share a common feature: they simplify everyday household tasks. The key is to formulate it so that a computer can do it. This process is called digitalization, and it represents the second stage of digital transformation (DX). The interview headline that appeared in the magazine was, "PCs can be so much fun at home." In today's terms, this could be called a life hack—a clever way to make daily life more comfortable, enjoyable, and efficient.

The next time cooking and computers came into my life was when I decided to study "learning in everyday life". I thought that metacognition was essential for proficiency in cooking. Metacognition means looking at one's intellectual activities (cognition) from a higher level. To devise intellectual activities, it is necessary to take a broader perspective. Metacognition is the ability to view intellectual activities objectively and adjust one's behavior.

In general, cooking is not learned at school but is often learned at home as an apprentice or by imitating others. We wanted to use computers to support this process of mastery. To develop a system to support proficiency in cooking, we began by observing the differences between novices and experts. We prepared a recipe for cooking two dishes simultaneously. Using the same kitchen system, we recorded the process on video. The difference between the novice and expert cooks was seen in the difference in cooking time. The average cooking time for the novices was 69 min, and the average for the experts was 37 min. The average cleanup time for the novices was 16 min, and surprisingly, for the experts, it was only 2 min. The expert cooks were almost done cleaning up as soon as they finished cooking.

If you watch the video of the cooking process carefully, you will see that expert cooks start by arranging the seasonings and utensils in a convenient place and setting up the work area. They also wash the colander and bowl after each use and keep the vegetable peels and other debris from the cooking process in one place (Fig. 4.5). On the other hand, novice cooks left the bowls and seasonings after using them and looked for where they were every time they needed them. This led to differences in cooking time and cleanup time. Experts could plan, or set up, the spatial and temporal homogeneity of their tasks.

In the cooking table of the novice in Fig. 4.5, the cooking oil is placed horizontally on top of miso. On the other hand, the tools and seasonings are placed in an organized

**Fig. 4.5** The difference between a novice and an expert in cooking

manner in the expert's kitchen. In the sink of the novice, the used items are placed haphazardly, while in the sink of the expert, the vegetable trash is placed in a group on the right side.

This is not the only ability needed while cooking. For example, if you want to cook three dishes using two stoves, you want each dish to be at the right temperature and on the table at about the same time. We want to minimize the cooking time while minimizing the number of dishes to be washed. Trying to solve a problem under such conditions needs computational thinking.

Nowadays, there are many recipes on the Internet, so the way of learning has changed. However, since these are single-item recipes, it is up to the individual to accumulate them as meta-knowledge. Here, meta-knowledge refers to transferable and versatile knowledge, such as "If you can make a good Oyako-don1F1F[2] (chicken and egg bowl), you can also make a good Tanin-don2F2F,[3]" or "When stewing, adding sugar before salt enhances the flavor." To create meta-knowledge, you may find commonalities, similarities, patterns, or even modularize.

Metacognition is also important during cooking. Metacognition allows us to compare the cooking process with the image of the finished product, taste it, and make adjustments such as turning down the heat on the stove. Furthermore, if we consider the cooking process, we can see three essential components of an algorithm: sequence, selection, and repetition.

Many cooking recipe books exist for different readers, such as novice, intermediate, and advanced. Nowadays, many recipes are on the Internet besides books. Moreover, in the COVID-19 pandemic, more and more people are cooking for themselves, and videos of chefs from famous restaurants appear.

Cooking and computational thinking are a perfect combination. It should be possible to design a new kind of recipe—a "computational thinking recipe"—that incorporates meta-knowledge, pattern recognition, metacognition, algorithms, and optimization as practical "tricks" that can be applied.

## 4.5 New Learning and Computer Science

The discussion so far is based on my 2017 colloquium presentation. Now, I would like to consider the new knowledge and skills arising.

At the end of the twentieth century, the knowledge and skills needed by children in the twenty-first century have changed dramatically. The Organisation for Economic Co-operation and Development (OECD) has published a new set of internationally agreed-upon key competencies, along with their assessment methods. Along with these key competencies, the importance of computer science as a learning content has also begun to be recognized. The field of educational technology has been classifying educational goals (cognitive, motor, emotional) since the 1950s, and what

---

[2] "oya"=parent=chicken, "ko"=child=egg.

[3] Unlike Oyako-don, Tanin-don uses eggs of a different species.

learners need to acquire has shifted from knowledge (what they know) to cognitive abilities (knowledge and skills), non-cognitive abilities (attitudes and values), and key competencies (what they can do).

The OECD has identified the achievement of these goals for education as a critical challenge for all countries and has been working on the Education 2030 project since 2015. In 2019, the OECD released a concept note as one of the final reports summarizing the project's results so far (Phase 1). The concept note adds a new category of transformative competencies to the existing categories of key competencies. The transformative competencies consist of the following three elements:

1. The ability to create new value: collaborating with others to develop new products, services, methodologies, ways of thinking, and new social models.
2. The ability to reconcile tensions and dilemmas: mediate between diverse ideas and interests and balance various competing demands.
3. Taking responsibility: to consider the future consequences of one's actions, evaluate risks and rewards, and take responsibility for the outcomes of one's work.

The emphasis here is that educating young people means preparing them for work and equipping them with the skills to become responsible, engaged citizens. It also emphasizes the importance of learner agency and the need for learning through anticipation, action, and reflection concerning the real world.

In contrast, school education in Europe and the U.S. introduces computer science and computational thinking into the curriculum from elementary school onward, as the twenty-first century is the age of computation. In the U.K., the national curriculum was revised in 2012 to include computer science at all levels, from 5 years old to elementary, middle, and high school, starting the following year. In the U.S., in a 2016 Executive Order, President Obama stated that all children from kindergarten to high school should learn computer science and that everyone should be able to think computationally. In response, computer science and computational thinking, together with STEAM (Science, Technology, Engineering, Art, and Mathematics) education, have been incorporated into curricula from elementary through high school.

By contrast, "programming-like thinking," a term widely used in the Japanese educational community, should not be confused with computational thinking, as the two are conceptually distinct. In programming thinking, the metaphor of "thinking like a computer" is often used. On the other hand, computational thinking is "thinking like a computer scientist," as I have explained so far. Computational thinking is a highly versatile approach to problem-solving that can be used for issues society is facing, such as climate change, genome analysis, and fake news detection.

In Japan, competencies are often translated as "qualities and abilities." Dr. Yutaka Sayeki, an educationalist and cognitive psychologist, points out that a competency defines what it means to be "competent," rather than being the motive or "power" that causes it. In other words, competency is not a quality or an ability but a capability that emerges in activities. If you read the original English text where the concept of competency was introduced, you will find the following: The ability to use language, symbols, and text interactively.

Then we are left with the question of how to develop such an ability (educationally speaking). However, Sayeki argues that if we understand it as "the ability to use language, symbols, and text interactively in various situations," then there would be no objection to it being one of the indicators of competency.

Sayeki cites "Ghost in the Machine" dogma in the book "The Concept of Mind" by philosopher Gilbert Ryle. This dogma is the assumption of the "mind" of the person that makes "intellectual behavior" possible. The misunderstanding lies in assuming that there is a cause called "mind" inside the "smart-acting machine" that produces "intelligent behavior". Rather, the "mind" is the sum of the possibilities (what Ryle calls "dispositions") for people's behavior to exhibit certain aspects.

Dr. Minoru Murai, a philosopher of education, explains the symptomatic principle of education in his book, "An Encouragement of New Pedagogy," using the parable of the quack who catches a cold. When Sayeki told him about this parable, he thought it was an interesting thing to say. The story is as follows:

> There was a country where a good cold catcher was considered a good person. All the parents wanted their children to be good cold catchers. At a meeting of professional doctors, a scientific study was conducted on the characteristics of ministers, presidents, rich people, and other people who were considered to have a good cold. The results showed that a person must have 1: a fever of 37 degrees or higher, 2: a headache, and 3: a feeling of dullness. To induce a fever of 37 degrees Celsius or higher, curry powder and horseradish must be kneaded together and rubbed all over the body. To induce a headache, they struck the head. To make them feel dull, have them carry a bale of rice around the playground three times. All over the country, this method was used on children. To see the effects of the treatment, tests were conducted by taking their body temperature. The result was a child with all three symptoms of a "full-blown cold.

Murai pointed out that what is being done in schools today is the same as this strange cold catching. What is so funny about this story? Where did the people of this country go wrong?

The story of the "quack who made people catch colds" parallels the tendency to treat competency as a fixed quality or ability to be acquired. The idea of goal-based evaluation comes from the fact that the goal is to develop the competencies and then to "evaluate" whether the goal has been achieved or not. Competency in education was initially discussed by the OECD in the Definition and Selection of Competencies (DeSeCo) project and appeared in the final report issued in 2003. In defining (clarifying) and extracting the competencies, elements of people who have shown competence in various industries were collected and organized. This approach is symptomatic of "catching a cold."

In other words, competency is a point of view, an indicator that we look at when we carefully observe and find the competencies that emerge. It is a shift to discovering the diversity of competencies that appear in daily life and work, finding previously unrecognized "goodness," recognizing it, paying renewed attention to its aspects, recognizing it in oneself and others, and developing it.

There is a shift from the traditional view of learning, where knowledge is accumulated, and skills are acquired, to a new view of learning, where knowledge is created through dialogue and competence is recognized in the process. As the view of learning changes, it is a natural consequence that the view of assessment and assessment methods will also change. Based on the competency discussion above, it is necessary to shift from the conventional evaluation of "whether or not the student has mastered the skill" to a new assessment of "whether or not the student is aware of the good points and is trying to improve." When we consider this competency discussion in conjunction with computational thinking, many things become clear. If we think of the elements of computational thinking described in the previous chapters as "the ability to do X," then the question becomes, "How can we educate people to do X?

In general, experts have a lot of useful knowledge and skills that they use without being aware of it. In most cases, they are difficult to define or explain in words. We know how to use many things in our field of expertise but cannot present in terms to non-specialists. This book boldly challenges that difficulty. Computational thinking is a skill that computer scientists display, and it is an approach to problem-solving that they use consciously and sometimes unconsciously. It is helpful for children, and when we try to teach it, we need to be very careful not to make the same arguments for competency that we have discussed so far. A competent person has X and Y elements when looking at an activity. To be competent, one needs to acquire elements X and Y. But if we stop there, we fall into what might be called the symptomaticism of quackery. So, let me explain.

A competent person has acquired the elements X and Y through a particular series of activities. If we think of it as "a series of activities," we can see a hint in performing and experiencing these activities. In addition, it is not just one person who conducts these activities, but also the existence of others, a community, that recognizes them as "good". The self, others, and the community reflect on the action results. If we assume that the characteristics of computational thinking are reinforced through such activities, then acquiring it requires designing an environment where learners can manifest, recognize, and develop their abilities by engaging with the world around them.

## 4.6 Computational Thinking for All

Finally, I would like to reflect on how computational thinking can be developed and applied.

In the discussion of competency, I mentioned that it involves discovering the diverse abilities that emerge in society—finding forms of "goodness" that had previously gone unnoticed, recognizing them, attending to them again, identifying them in oneself and others, and nurturing them. Consider computational thinking as one of those perspectives of "goodness".

For example, if you want to make many cookies with a heart-shaped pattern in the center (Fig. 4.6), you could make small pieces of chocolate cookie dough, each shaped like a heart. Then, combine each heart-shaped piece with butter cookie dough and bake them together.

**Fig. 4.6** Heart cookies

Some of you may have imagined a more efficient way: shaping the chocolate dough into a rod with a heart-shaped cross-section, wrapping it with butter cookie dough, and slicing it into pieces. This is how to make Kintaro candy or the "decorative sushi rolls" that have recently become a bit of a craze. The latter method requires less work and is also more consistent in size.

Then, say aloud.

"Isn't this computational thinking? Yes, it is."

"Isn't this similar to the way you make a beautiful, artistic terrine?"

At some point, you may suddenly realize that this is computational thinking. You may also realize it when you see what others are doing. It's okay if you're wrong. Let's put it into words first. Then, let's say it out loud. After doing this a few times, you may suddenly realize, "Oh, I'm learning computational thinking." Or someone might even tell you, "You're a person who can think computationally."

In the 1980s, many educational studies using LOGO revealed that programming skills alone did not necessarily lead to the development of logical thinking or problem-solving abilities. Subsequent research in learning science has revealed that learning transfer (the influence of previous learning on later learning) does not occur automatically across contexts but is context-dependent and involves experience. This is one of the reasons why Project-Based Learning (PBL) has become an effective learning method in the world today. Therefore, computational thinking can also be practical if used in activities such as project-based learning. In programming and STEAM education activities, social activities, and co-creative design activities, we can design situations where computational thinking can emerge, find it, and develop it.

By adding or limiting the characteristics of computational thinking to competency "the ability to use language, symbols, and text concerning each other in various situations," we may be able to create an index of competency in computational thinking. However, this will require further discussion.

Finally, I would like to emphasize that when we think about teaching and learning computational thinking, we should avoid looking for "ghosts" or collecting "symptoms." Instead, we should weave computational thinking naturally into our daily lives, so that many people can make use of it in practice. Above all, I think it is essential for everyone to enjoy executing computational thinking.

Let's make computational thinking—once enjoyed mainly by computer scientists—accessible to everyone. **Computational Thinking for All!**

**Column 11 Calculation and computation, two concepts of computing**

The term computation existed long before the invention of computers. Today, calculation and computation are often used interchangeably, but this was not always the case.

Charles Steinmetz, a mathematician and electrical engineer, and Vannevar Bush, who proposed the Memex information retrieval system, regarded calculation and computation as key concepts in electrical engineering and believed that the computing revolution had begun even before the 1940s. They adopted these two concepts in engineering textbooks around the same time to distinguish between analyses requiring high levels of technical sophistication and those that did not, as well as between low and high degrees of mechanization.

(*Source* Life Magazine, September 10, 1945, Vol. 19, No. 11, p. 123)

The word computer has a broader meaning than the modern sense of a digital machine. From ancient carved stones and counting pebbles or beads to the abacus and slide rule, humanity's "tools for counting" are regarded as the ancestors of modern digital computers. Other examples include calendars, sundials, globes of the solar system, astrolabes (astronomical instruments used by ancient astronomers and astrologers), planetariums, and various physical or conceptual models such as scale models and mathematical models.

From the Renaissance until the advent of the electronic computer, a computer referred to a "computing person" or someone in that profession (see Column 7). During World War II, the demand for ballistics tables exceeded what could be calculated by hand, leading to the development of ENIAC—a computer capable of solving a wide range of computational problems at high speed. Subsequently, computers were mass-produced as hardware costs decreased and electronic components became smaller. In the 1990s, the Internet and the World Wide Web, where computer networks are interconnected on a global scale, became widespread, and microprocessors were embedded everywhere, from home appliances to automobiles. Microprocessors became ubiquitous—from home appliances to automobiles—handling massive amounts of data and solving a wide variety of problems. As computers permeated society from the 1950s onward, they freed humans from the labor of patiently performing simple calculations, and the word computer came to denote a machine rather than a person.

The word calculate, on the other hand, now refers to determining a specific quantity or value using arithmetic operations. Like computers, calculators have evolved with technological development—from simple counting tools to electronic desktop devices. In the nineteenth century, mechanical cash registers were invented, allowing simple transactions such as store payments to be performed by machines. In the mid-1960s, calculators partially employing ICs appeared. In the early 1970s, products appeared with names such as "electronic slide rule" and "electronic soroban." In addition to specialized office calculators that could perform four arithmetic operations and percentage calculations, function calculators, programmable calculators, and graphing calculators also appeared, but their use has become limited as computers have become more widespread. Today, calculators are embedded in application software for cell phones and smart phones and continue to be used by people.

Based on these historical developments, while calculation and computation both essentially mean "computation," calculation today refers primarily to the use of mathematical formulas, especially the four arithmetic operations. Computation, by contrast, encompasses the systematic application of advanced mathematical rules and the processing of large amounts of data. It is thought that computational thinking emerged from efforts to conceptualize "computer-assisted computation" and to explore its broader possibilities.

# Chapter 5
# Joy of Understanding Computational Thinking

**Misako Nambu**

**Abstract** In this chapter, the author, a cognitive scientist with no background in computer science, shares her experience with computational thinking. She describes her initial discomfort with the concept and the journey she took to fully embrace it. As mentioned in the Preface, many people struggle to help beginners and non-experts grasp the essence of computational thinking. This chapter and this book aim to offer a helpful guide for those individuals. The author begins by analyzing her initial discomfort, the purpose of learning computational thinking, and the mindset needed to approach it. In the midst of this process, a simple comment from a colleague became a turning point, offering her a glimpse into the benefits of computational thinking and ultimately changing her entire perspective. Finally, she discusses the unique characteristics of computational thinking she discovered as a non-specialist, focusing specifically on its altruistic power.

## 5.1 A Non-computer Science Expert's Journey into Computational Thinking

Many of the readers of this book are not experts in computer science. I am a faculty member at our university (Future University Hakodate; FUN), but my expertise is in cognitive science, not computer science.

As described in the Preface, when this computational-thinking book project started within FUN, I witnessed firsthand how faculty members who specialize in computer science deeply engaged with computational thinking.

However, when I read Wing's paper, I could understand in theory the benefits computational thinking could bring to humans and the world, but I was not as convinced as the computer science experts were. What specific aspects of computational thinking are so inspiring or exciting to computer science experts? This leads

M. Nambu (✉)
Future University Hakodate, Hakodate, Hokkaido, Japan
e-mail: m-nambu@fun.ac.jp

 
H. Nakashima and K. Hirata (eds.), *Computational Thinking*,
https://doi.org/10.1007/978-981-95-5962-6_5

to the question: What is the difference between their understanding of computational thinking and non-experts' perception of it, and what might explain this gap?

In this chapter, I would like to consider that question from the perspective of a non-expert in order to promote computational thinking more widely throughout the world.

## 5.2 Wing's Computational Thinking

Let's delve into what a well-known essay on computational thinking suggests. When I first read it (in translation), these points stood out:

- Computational thinking is a general attitude and skill
- It is basic technology for everyone
- It should be taught alongside reading, writing, and arithmetic
- It means thinking like a computer scientist
- It involves understanding the capabilities and limitations of computational processes, whether the agent is human or machine
- It includes judging not only correctness and efficiency, but also aesthetics and elegance in system design
- It means transforming complex problems into ones we already know how to solve
- It gives us the courage to tackle problems and build systems that we couldn't solve alone

My honest first impression was, "So, what exactly should I do?" The essay laid out a vision from a broad perspective, but it felt abstract and hard to grasp. Terms like "aesthetics" and "elegance" are difficult to define and unfamiliar to me.

Some colleagues in computer science often say, "I'm not sure what a good example of computational thinking is" or "We're still struggling to articulate the benefits of computational thinking." It reminded me of how people can ride a bicycle quite naturally, even if they can't explain the mechanics behind it. It's something done naturally and unconsciously. That's why I think a non-expert like me might offer a unique perspective in exploring this topic.

## 5.3 Living Better and Happier

Based on what I've learned so far, computational thinking seems to involve:

1. Understanding the complexity of the world through computation (description)
2. Solving problems (processing)
3. Living better and happier (goal)

The first two points are procedural steps toward achieving the third. That's the tricky but fascinating part—how computational thinking can contribute to a better quality of life.

Even with this framing, uncertainties remain. If being a computer scientist means fully understanding and practicing the first two, do we all need to go back to school? I don't think it is very desirable to draw a line between experts and non-experts in computer science. However, the gap can feel vast. How do we bridge it?

One of the authors asked, "Do we even share the same goal in the first place?" That is, is our idea of "living better and being happy" the same as theirs? It's hard to answer, but I hope that at least the goal is shared, even if the paths are different. Otherwise, there's no point in having this dialogue.

The same person also asked whether non-experts can really grasp ideas like "aesthetics" and "elegant system design" that computer scientists often use. I think non-experts can. These ideas may manifest differently, but there are shared values across disciplines. We may be using different words, but we're all reaching for some kind of aesthetic understanding—beyond just efficiency.

In this sense, even without technical knowledge, people are already thinking computationally in how they observe and solve problems, collaborate, and try to build a better society. This thinking is not just about skills or knowledge, but also attitude and that opens space for dialogue, even about concepts like "beauty."

## 5.4 Asking the Unasked: The Path to Shared Understanding

As I said before, I wasn't sure why I was invited to be part of this book project. I used to conduct quantitative psychological experiments, but now I mostly do qualitative research and fieldwork, focusing on designing communication environments. In any case, I'm not a computer scientist.

Looking back, though, having someone trying to understand—even imperfectly—can challenge assumptions that experts take for granted. I was frequently asked, "Did you understand that?" at our sessions. At first, I was nervous—worried my ignorance would be exposed. But over time, I realized that the others weren't being critical—they were genuinely enjoying the dialogue. It reminded me of the deep conversations I used to have in graduate school.

There's a study I want to mention. In science museums, communicators are advised not to say, "Are you familiar with this?" to visitors—it's considered taboo. But analysis of real interactions shows that, in context, such questions can actually trigger engagement and two-way communication. Likewise, when I was asked, "Did you understand?" I think it was less about testing my knowledge and more an invitation: "Let's think together."

## 5.5 The Moment Computational Thinking Finally Clicked

Wing's essay mentioned earlier lists many technical elements of computational thinking, including:

- instruction sets, computational resources, operational constraints
- optimization, approximation, randomization
- recursion, simulation, abstraction, transformation
- error correction, redundancy, deadlock, caching
- heuristic reasoning, type checking, procedure calls

Reading this kind of list, it's easy to forget the broader goals like "living better" or "creating a better society." These terms may not sound exciting to non-experts, but if we understand how they relate to our own lives, maybe they become more meaningful.

One day, I tweeted something like: "Help! Tasks keep piling up. I'm overwhelmed. Breathe, breathe…" A computer scientist calmly replied, "In computer science, we call that a *stack overflow*."

That moment stayed with me. I would have just called it chaos—but now I had a technical name for it. I looked it up and learned that it refers to too many function calls in a program, often caused by recursion or large data structures. It suddenly made sense. I could describe and analyze my mental state using this concept and even consider how to solve it. It might sound exaggerated—but it was a vivid experience.

So again, computational thinking isn't limited to computer scientists. We all engage in it, often unconsciously. With help from computer science concepts, we can reflect on it more clearly and improve how we solve problems in everyday life.

Reflection itself is not a solitary act: by sharing the language and metaphors of computation, we turn inward insights into an invitation to others. In this way, reflection becomes altruistic because it naturally leads to thinking together—a movement from the self toward dialogue and collaboration.

## 5.6 Thinking Together, Building Better: The Altruistic Power of Computational Thinking

Let me summarize what I've come to understand:

- Computational thinking is more than knowledge or procedures
- It's about what we do with them
- It sharpens observation, enhances analysis, and aids problem-solving
- It enables dialogue around concepts like beauty and goodness
- Ultimately, it helps us live better and create a better society

Understanding computational thinking is, indeed, a pleasure. The knowledge of computer science becomes a powerful tool for thought and connection. Even for non-experts, exploring this world is a profound experience. I now see why experts are so passionate about it.

In today's world full of complex challenges, computational thinking is more relevant than ever. It's not about individual gain, it's about working together with diverse people to solve societal problems. It provides a foundation for working together with others—a kind of altruism rooted in collaboration. As one philosopher said, good altruism always involves discovering others and transforming oneself. That's what this dialogue felt like: discovering, changing, and growing with others.